Practical Model-Driven Enterprise Architecture

Design a mature enterprise architecture repository using Sparx Systems Enterprise Architect and ArchiMate® 3.1

Mudar Bahri

Joe Williams

BIRMINGHAM—MUMBAI

Practical Model-Driven Enterprise Architecture

Associate Group Product Manager: Richa Tripathi
Publishing Product Manager: Shweta Bairoliya
Senior Editor: Nisha Cleetus
Content Development Editor: Rosal Colaco
Technical Editor: Pradeep Sahu
Copy Editor: Safis Editing
Project Coordinator: Manisha Singh
Proofreader: Safis Editing
Indexer: Pratik Shirodkar
Production Designer: Prashant Ghare
Marketing Coordinator: Deepak Kumar

First published: April 2022
Production reference: 1250422

Published by Packt Publishing Ltd.
Livery Place
35 Livery Street
Birmingham
B3 2PB, UK.

ISBN 978-1-80107-616-6

www.packt.com

To my mother – thank you for your prayers. To my children, Samer, Lina, and Lara – believe in yourselves!

– Mudar Bahri

To my wife, Robin – you are the love of my life. To my son, Ryan, and daughter, Kayla, because I'm so proud of you both. To my grandson, Ethan, because I adore you.

– Joe Williams

Contributors

About the authors

Mudar Bahri is an enterprise architect with long, progressive experience in implementing **TOGAF®** in large and midsized organizations. He is an independent **Enterprise Architecture** (**EA**) mentor and consultant and helps architects to build and develop practical EA skills that can provide decision-makers with useful artifacts. He is a certified TOGAF® 9.1 Enterprise Architect with experience in other EA frameworks and project management methodologies. He has used TOGAF® to develop digital transformation strategies, cloud migration plans, solution architectures, and proposals. He believes that EA is meant to bridge gaps, not create new ones, so it has to be simple, practical, and serve everyone.

> *I want to thank the people who have been close to me and supported me,*
> *especially my brother, Tameem Bahri, who inspired me to start writing.*

Joe Williams has over 40 years of experience in software engineering and architecture on many platforms and spanning several business domains, including retail, banking, health insurance, telecommunications, environment monitoring, and government systems. He holds a BS degree in information systems from the University of San Francisco and is a TOGAF® 9.1 certified practitioner. Joe preaches the practice of modeling every chance he gets. He is an expert in the **Unified Modeling Language** (**UML**) and has helped establish architecture practices at several organizations. He has been modeling solutions using Sparx Systems Enterprise Architect since 2007 and other modeling tools since 1995. He is currently retired and living in northern California.

> *I want to thank my co-author, Mudar, for asking me to help write this book,*
> *and my wife, Robin, for encouraging me to do so.*

About the reviewer

Kris Marshall is a results-driven chief enterprise architect with over 20 years of accomplishments in information technology leadership, enterprise architecture, learning technologies (artificial intelligence and machine learning), and organizational transformation. Kris has extensive experience with business strategy development (for over 35 public and private organizations) and defines business and solution architectures in support of those strategies, aligning organizational goals and the IT infrastructure that support them.

As a doctoral graduate of Pepperdine University, Kris' research focuses on enterprise architecture theory and the intersectionality of learning technologies. Kris presently supports numerous university research labs and participates in research groups at Pepperdine University, MIT, and Singularity University.

Table of Contents

Section 2: Building the Enterprise Architecture Repository

3

Kick-Starting Your Enterprise Architecture Repository

4

Maintaining Quality and Consistency in the Repository

5

Advanced Application Architecture Modeling

6

Modeling in the Technology Layer

7

Enterprise-Level Technology Architecture Models

8

Business Architecture Models

9

Modeling Strategy and Implementation

Section 3: Managing the Repository

10

Operating the EA Repository

11

Publishing Model Content

Index

Other Books You May Enjoy

Preface

Enterprise Architecture (**EA**) is a discipline that promises to bridge the gaps between business and IT by defining the components of an enterprise and also the relationships among these components. With the increased dependency of businesses of all domains on technology, many organizations want to adopt EA, and most professionals want to be part of this adoption process. However, following a discipline at the entire enterprise level can be full of obstacles because it requires dealing with many people with different interests. This is where most EA practitioners struggle, and they are perceived as theoretical people who are disconnected from reality; as a result, many organizations lose interest in EA.

In our opinion, the lack of delivering tangible and useful EA artifacts is the main reason for EA implementation struggles. This book introduces sample artifacts on each element in an enterprise so that you are able to address the different needs of different stakeholders. A collection of EA artifacts forms an EA repository, so this book will teach you how to build a repository and maintain its content. It will also teach you how to build a reference library that you and your EA team can use to maintain consistency within the EA repository.

We chose ArchiMate® 3.1 to build these artifacts, but you can use any other modeling notation that works within your workplace. We also chose Sparx Systems' Enterprise Architect software as the tool to build and develop these EA artifacts, but you're also free to use any other tool that you are familiar with. The goal is to deliver value, regardless of the tool that you prefer to use.

Who this book is for

This book is for enterprise architects who are charged with influencing the business, technology, and applications of an organization. The authors are influenced by TOGAF®. Whether you work on the business, applications, data, or technology layer, or all of them, this book covers examples that you'll be able to apply in your work. Experience in EA frameworks, especially TOGAF®, is beneficial, but not required. Although not necessary, familiarity with modeling with Sparx Systems Enterprise Architect using any modeling language will be helpful. No prior knowledge of ArchiMate® is required to get started with this book.

What this book covers

Chapter 1, *Enterprise Architecture and Its Practicality*, starts by highlighting what made TOGAF® the de facto standard for implementing EA and puts the spotlight on the problems that most TOGAF® practitioners face – some (if not all) of which we are quite sure you will have faced. In this chapter, we will talk about the problems, the proposed solutions, and the tool that will be used for this purpose.

Chapter 2, *Introducing the Practice Scenarios*, introduces the three scenarios that will be used to make this book practical and connected to real-life examples. The company that we will talk about is a fictional company, which we will name ABC Trading, but it can represent any company that you have or will be working for. The purpose of the scenarios is to make up simple and generic stories that can fit within your real-life work environment, even if your organization is working in a different business domain. The scenarios will be detailed enough and as close as possible to problems that you have faced or will face as an EA.

Chapter 3, *Kick-Starting Your Enterprise Architecture Repository*, helps you create your first diagram instruction by instruction to get you familiar with Sparx. By doing so, you will be introduced to some techniques at different levels of expertise that you may find useful, even if you have used Sparx before.

Chapter 4, *Maintaining Quality and Consistency in the Repository*, helps you learn how to use the ArchiMate® 3.1 standard as a reference by putting it in a simple and easy-to-understand way. Therefore, architects contributing to the content of the repository will enjoy following the standard, which will be reflected in the quality and consistency of their artifacts.

Chapter 5, *Advanced Application Architecture Modeling*, helps you learn how to enrich the EA repository with additional artifacts from the application architecture layer. You will learn more advanced skills in Sparx and ArchiMate® to build an EA repository that can be trusted within your organization.

Chapter 6, Modeling in the Technology Layer, introduces you to the structural elements of the technology layer and how they are related to other layers. We will learn how to model technical, physical, and network environments. We will also learn different ways to represent technical constructs and how to move from more abstract to more concrete representations.

Chapter 7, Enterprise-Level Technology Architecture Models, introduces behavioral elements and how to use them. We will learn how to import information from other sources and model and organize technical services in order to answer enterprise-wide questions. We will generate our first report and analyze our findings.

Chapter 8, Business Architecture Models, helps you answer business questions, such as what a business provides to the world, how it achieves it, how an organization is structured, who is responsible for what, and how these business architecture elements are automated and realized by elements from the application and technology layers.

Chapter 9, Modeling Strategy and Implementation, helps you learn how to utilize the power of business capability models to identify the gaps between what we are capable of doing now and what we are targeting to do in the future. Once the gaps are defined, you will need to put a plan for bridging them within a timeline.

Chapter 10, Operating the EA Repository, helps you learn how to add formality and organization to the process of making changes to the enterprise repository. The more content you have and the more architects add and remove content to and from the repository, the more challenging it becomes to keep it organized and up to date.

Chapter 11, Publishing Model Content, helps you learn how to put diagrams and content that you have developed in the previous chapters into documents, presentations, or web pages that can be shared with other stakeholders who do not have experience with Sparx or a license to use it.

To get the most out of this book

Here are the technical requirements for this book:

Software/hardware covered in the book	Operating system requirements
Sparx Systems Enterprise Architect 15.2	Windows, macOS, or Linux

If you are using the digital version of this book, we advise you to type the code yourself or access the code from the book's GitHub repository (a link is available in the next section). Doing so will help you avoid any potential errors related to the copying and pasting of code.

Portions of this text are reprinted and reproduced in electronic form from The Open Group® ArchiMate® 3.1 Specification with permission granted by The Open Group®, L.L.C. In the event of any discrepancy between text and the official standard, the official standard found at `https://www.opengroup.org/library/c197` remains the authoritative version for all purposes. ©The Open Group. All rights reserved.

Download the example repository files

You can download the example EA repository files for this book from GitHub at `https://github.com/PacktPublishing/Practical-Model-Driven-Enterprise-Architecture`. If there's an update to these files, they will be updated in the GitHub repository.

We also have other code bundles from our rich catalog of books and videos available at `https://github.com/PacktPublishing/`. Check them out!

Download the color images

We also provide a PDF file that has color images of the screenshots and diagrams used in this book. You can download it here: `https://static.packt-cdn.com/downloads/9781801076166_ColorImages.pdf`.

Conventions used

There are a number of text conventions used throughout this book.

`Code in text`: Indicates code words in text, database table names, folder names, filenames, file extensions, pathnames, dummy URLs, user input, and Twitter handles. Here is an example: "Create a new ArchiMate® 3.1 application diagram in the Metamodels package and name it `Application Process Focused Metamodel`."

Bold: Indicates a new term, an important word, or words that you see onscreen. For instance, words in menus or dialog boxes appear in **bold**. Here is an example: "Right-click on the **Manipulate Device Location Data** application process element."

Tips or Important Notes
Appear like this.

Get in touch

Feedback from our readers is always welcome.

General feedback: If you have questions about any aspect of this book, email us at customercare@packtpub.com and mention the book title in the subject of your message.

Errata: Although we have taken every care to ensure the accuracy of our content, mistakes do happen. If you have found a mistake in this book, we would be grateful if you would report this to us. Please visit www.packtpub.com/support/errata and fill in the form.

Piracy: If you come across any illegal copies of our works in any form on the internet, we would be grateful if you would provide us with the location address or website name. Please contact us at copyright@packt.com with a link to the material.

If you are interested in becoming an author: If there is a topic that you have expertise in and you are interested in either writing or contributing to a book, please visit authors.packtpub.com.

Share Your Thoughts

Once you've read *Practical Model-Driven Enterprise Architecture*, we'd love to hear your thoughts! Scan the QR code below to go straight to the Amazon review page for this book and share your feedback.

https://packt.link/r/1-801-07616-2

Your review is important to us and the tech community and will help us make sure we're delivering excellent quality content.

Section 1: Enterprise Architecture with Sparx Enterprise Architect

This section briefly introduces **Enterprise Architecture** (**EA**) and TOGAF®, and explains why this book can be of value to you and how we have structured it to be easy to follow.

This section addresses the problems that are faced by EA practitioners and introduces you to the methodology, the modeling notation, and the modeling tool that will be used as components of a solution. It addresses why some organizations have lost interest in EA, but more importantly, it introduces solutions to regain business trust in it again.

Since this book is mainly concerned with making EA practical, we will introduce three fictional problem scenarios that can guide you to create useful artifacts and show some tangible values of EA in your workplace.

This section comprises the following chapters:

- *Chapter 1, Enterprise Architecture and Its Practicality*
- *Chapter 2, Introducing the Practice Scenarios*

1
Enterprise Architecture and Its Practicality

Enterprise Architecture (**EA**) is a discipline that many organizations have adopted or have been motivated to adopt over the last two decades or so due to its promises to bridge the gaps between business and technology. EA is the art of defining and categorizing the elements that compose an enterprise and defining the relationships among these elements to get useful information that supports making strategic and tactical decisions. There are several frameworks that guide EA implementation, but the most popular one is **TOGAF®**.

This chapter starts by highlighting what made TOGAF® the de facto standard for implementing EA and puts the spotlight on the problems that most TOGAF® practitioners face – some (if not all) of which I am quite sure you will have faced. As I have learned, talking about problems is never helpful without providing solutions, so we will introduce a hands-on approach that has been extracted from years of practical experience in the EA domain to help you in aligning the theory with the practice smoothly and more productively.

Please remember that this book is not about teaching TOGAF®; I expect that you already have some knowledge of and experience with the framework and are looking for solutions to the problems that you may have already faced. It is also not about making comparisons between TOGAF® and other frameworks to show the advantages versus disadvantages of each. This book is based on TOGAF® and ArchiMate® only and will explain how to use them in a way that can help your organization to get quick, tangible outcomes from adopting them.

The following is a list of topics that will be covered in this chapter:

- Understanding TOGAF®
- Introducing agile EA
- Introducing ArchiMate®
- Introducing Sparx Systems Enterprise Architect

Let's start by talking about the benefits and drawbacks of TOGAF®.

Understanding TOGAF®

Even though TOGAF® came nearly two decades after the **Zachman Framework** (`https://www.zachman.com/about-the-zachman-framework`) was introduced, it dominated the market very quickly and became one of the most important standards in the EA domain. John Zachman was the first to introduce the concept of EA in the mid-eighties and defined an EA framework carrying his name. For many reasons, the Zachman Framework was not adopted by many architects, but the idea remained in many people's minds.

TOGAF® started to gain popularity in late 2002 when The Open Group® introduced version 8.0. From there onward, it continued to gain popularity and started to become the de facto standard in the EA domain especially when The Open Group® released version 9.0 in early 2009, followed by 9.1, and finally 9.2 in 2018. TOGAF® became popular because it provided enterprise architects with rich content that guides their development journeys and makes implementing EA achievable.

Architects chose to follow TOGAF® for many reasons, which we will talk about later in this section. However, implementing TOGAF® was not a straightforward journey for many, and it brought new challenges and difficulties to the architects. As a result, many EA projects ended up with massive scope creep, unneeded outcomes, and useless acronyms. Therefore, many EA projects got terminated due to low return on investment and more people lost faith in EA as a practical approach even with TOGAF®. In this section, we will talk about the following:

- The benefits of using TOGAF® as a framework for implementing EA projects
- The drawbacks that make implementing TOGAF® challenging

While you read this section, I am sure that you will recall similar situations that you or your team have faced in your EA implementation journey.

TOGAF® implementation benefits

The following features are some advantages that made TOGAF® the preferred choice over other frameworks for many architects:

- Complete online documentation that is freely available.
- An easy-to-follow process.
- It fits architects with different experience.
- A rich content metamodel.
- It's loaded with guidelines and techniques.
- It encourages learning.

We will look at each benefit individually in the following subsections and see why more than 111,000 individuals from over 144 countries have chosen to use TOGAF® and be certified in it (according to `https://togaf9-cert.opengroup.org/certified-individuals` on the date of writing this paragraph).

Complete online documentation that is freely available

The Open Group® has provided all the TOGAF® versions online and for free with anonymous access. This makes it possible for people at all levels of experience to explore, read, and learn the framework at their own pace without feeling constrained by costly subscriptions or time-limited trials. You do not even need to register on the website to be granted access to the content, which is something that not all frameworks provide. Some frameworks require paid memberships, and some require at least creating a profile, but this is not the case with TOGAF®. EA practitioners also find it very convenient to have the material online and accessible anytime, anywhere, and on any device.

> **Note**
> Even after being TOGAF® certified for years and practicing it continuously for about 15 years, I always have the website bookmarked in my browser.

An easy-to-follow process

One of the core TOGAF® components is the **Architecture Development Method (ADM)**, which is a series of phases, each with a defined set of inputs, steps, and outputs. Architects find it easy to follow the ADM, especially architects coming from an IT background. They all know that if you want to build a solution, you need to first envision it, define its requirements, plan it, design it, build it, deploy it, and then operate it, which is very well known as the **System Development Life Cycle (SDLC)**. The ADM has a similar concept to the SDLC, but the objective is to architect the entire enterprise and not a single IT solution.

The following diagram represents the ADM cycle as defined by TOGAF® here: `https://pubs.opengroup.org/architecture/togaf9-doc/arch/chap04.html`, using ArchiMate® 3.1 notation:

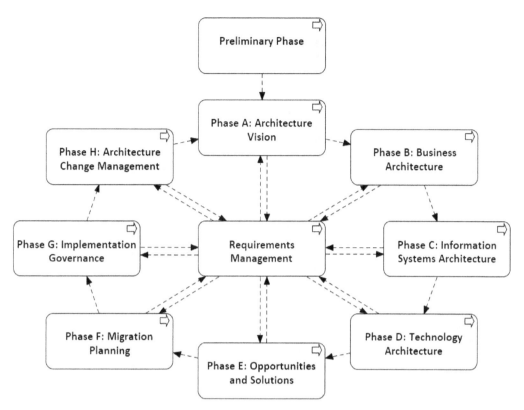

Figure 1.1 – An ArchiMate® representation of the ADM cycle (The Open Group®)

As you can see, the ADM cycle starts with the **Preliminary Phase**, which establishes the EA practice and formalizes the choice of tools and methodology. It is then followed by eight iterative phases that cover one part of the enterprise at a time.

It fits architects with different experience

Architects coming from different backgrounds and with different experience can all find something useful in TOGAF® that they can use in their area of expertise. Solution architects who have TOGAF® experience will have a better understanding of a business and be able to provide it with the right solutions that address its requirements and strategic directions. A second example would be project managers with TOGAF® experience, who will find it easy to upgrade their project management skills to program and portfolio management because understanding TOGAF® helps them understand how the enterprise works and how to have a holistic view of the different moving parts. IT operations managers with TOGAF® experience can use its **Technical Reference Model (TRM)** to categorize and classify the technology stack in their organizations using an industry-standard method, which helps them make the right decisions.

The list of examples can include every single member within the enterprise, so TOGAF® is an extremely useful and flexible framework that offers something for every practitioner, and each person will benefit from it differently.

A rich content metamodel

The **metamodel** provides architects with a foundation that tells them what the components of the enterprise are and how they are related to each other.

The TOGAF® metamodel, along with the taxonomy, provides architects with a better understanding so that they can describe the enterprise with consistency even if they look from different perspectives. Classifying and categorizing the elements and relationships of the enterprise is the heart and soul of EA. Here is the full TOGAF® 9.2 content metamodel: `https://pubs.opengroup.org/architecture/togaf9-doc/arch/chap30.html#tag_30_03`.

Additionally, TOGAF® provides a taxonomy and definition for all the keywords that architects use, which helps in unifying the language among different architects and makes their communications and documentation easier. People from different backgrounds can still argue the meanings of these definitions based on their interpretations but having a taxonomy can dramatically reduce these arguments. The Open Group® has even extended the TOGAF® taxonomy through ArchiMate®, which makes the combination of the two a complete and detailed set.

> **Important Note**
> This book will use the taxonomy defined by The Open Group® in both the TOGAF® 9.2 and ArchiMate® 3.1 materials as they are and will elaborate more for greater clarity.

It's loaded with guidelines and techniques

TOGAF® has tried to address all the issues that enterprise architects may encounter during their implementation engagements. It provides a set of useful tools, materials, and techniques, such as the following:

- A set of principles that enterprise architects can start with and modify as needed.
- Stakeholder management techniques and a proposed set of artifacts that can be of interest to different stakeholders.
- Patterns and techniques to segment and control the scope and size of the EA practice.
- A high-level governance model that can fit any size organization and team.
- A general structure of the EA content repository that can be scaled up or down as needed.

The complete list of guidelines and techniques is too long to be covered in this book and you may have already experienced some or all of them. Please refer to `https://pubs.opengroup.org/architecture/togaf92-doc/arch/chap17.html` if you want to learn more about them.

It encourages learning

With more people showing an interest in TOGAF®, The Open Group® wanted to encourage practitioners to be distinguished by becoming certified in the framework. With the increasing demand for experienced enterprise architects, becoming TOGAF® certified is a desire and sometimes a requirement by employers when hiring or contracting.

Just like anything good in life, the benefits of the framework come with a cost, which can outweigh the benefits if not handled with care. After having a quick look at the benefits of using TOGAF®, let's review some of the drawbacks that can be associated with TOGAF® implementations.

TOGAF® implementation drawbacks

Even with all the materials that TOGAF® offers, not every implementation project ends as planned, if it ends at all. Some EA implementations end up delivering something different than what was originally planned, some end up in a massive scope creep that overwhelms the project with infinite tasks and activities, and some remain within the drawers of the EA team until the sponsors decide to pull the plug.

In this section, we will highlight the things that may contribute to these results, and will provide you with some example traps that some architects may have fallen into; you may have witnessed or experienced some of them during your EA journey:

- The ADM is a giant waterfall process.

- The lack of practicality.

- High cost-to-value ratio.

- TOGAF® is mostly adopted by IT people.

Let's look at each of these in more detail in the following subsections.

The ADM is a giant waterfall process

As mentioned earlier, the ADM is a major factor in the success of TOGAF®. However, ADM is a sequence of phases, which turns out to be a waterfall method by nature. Therefore, TOGAF® provided a chapter explaining how to use the ADM iteratively (https://pubs.opengroup.org/architecture/togaf9-doc/arch/chap18.html), and another chapter on **Architecture Partitioning** to help keep the scope of work under control by dividing it into smaller partitions (https://pubs.opengroup.org/architecture/togaf9-doc/arch/chap36.html). Even with these techniques, there is still huge room for making scope mistakes because following a phased approach can end up with one phase requiring completion before starting the next. Attempting to finish a phase with all the details, inputs, steps, and outputs that are defined in the ADM can leave you and your team with an infinite set of tasks to do.

> Note
>
> Please keep in mind that I am talking from practical experience and how I have seen things ending up in most cases. The problem is not in TOGAF® itself, but with the wrong interpretation by the practitioners of how many details need to be defined for each ADM phase.

For example, in the ADM, architects start with the **Preliminary Phase**, which has an objective to *define and establish the detailed processes and resources for Architecture Governance* (https://pubs.opengroup.org/architecture/togaf9-doc/arch/chap05.html). Having this objective is understandable because, without governance, no project will be completed.

Establishing the **Architecture Governance Board** (**AGB**), which is the governance body that approves or rejects changes to the architecture, and defining their roles and responsibilities, their job descriptions, the governance processes, and the key performance indicators can take anywhere between a single day and 6 months, which is what I like to call *the Effort Blackhole*. This refers to a situation in which all the effort that you and your team put into finishing the phase will never end and there will always be the need for more.

Additionally, having this objective in the Preliminary Phase before the project is officially started makes it difficult for stakeholders and governance board members to understand how to govern something that does not yet exist. They can outline some processes and assign responsibilities, but that must start at a high level, then they'd need to get details as the organization's architecture maturity level increases. They may easily end up spending their time defining and refining these processes because it is not clear yet what to govern. The Effort Blackhole starts to form when the EA lead insists that all tasks in the Preliminary Phase must be completed before moving on to the next phase.

Another example is also in the Preliminary Phase of the ADM, which defines a *tailored architecture framework* as one of its outputs. A tailored architecture framework includes the following:

- A tailored architecture method
- Tailored architecture content (deliverables and artifacts)
- **Architecture principles**, including **business principles**
- Configured and deployed tools

That is according to *section 5.4* in *Chapter 5* in the TOGAF® online documentation (`https://pubs.opengroup.org/architecture/togaf9-doc/arch/chap05.html`). If this statement is not handled properly by the lead enterprise architect, it can result in a massive scope creep that can keep the entire team of architects busy for months in tailoring the architecture framework they plan to use. The EA team can easily end up building a new EA framework instead of following TOGAF® to deliver EA artifacts, that is, being trapped in another *Effort Blackhole*.

The last example I will mention in this context is in phase B (**Business Architecture**), which takes *Business Principles, Business Goals, and Business Drivers* as inputs (`https://pubs.opengroup.org/architecture/togaf9-doc/arch/chap07.html`). These elements can usually be found in the organization strategy, and if they are not, the architecture team will find themselves busy redefining (or just refining in the best case) the organization strategy, which is a project by itself and can lead to another massive scope creep and distraction to the team from their main EA objectives.

The lack of practicality

Another big problem that enterprise architects deal with when they plan to follow and implement TOGAF® is the lack of practical examples and document templates to use. Most frameworks impart theoretical information only to remain as generic and as neutral as possible. This leaves architects with huge gaps between the theory and the practice and in some cases, they deliver documents and presentations that are very theoretical and disconnected from the real world. This situation makes stakeholders perceive the EA office as the *ivory tower* office, indicating its disconnect from reality.

TOGAF® is full of terminologies that are either all new to many stakeholders or are defined slightly differently from other standards. For example, the definitions of *object* and *service* according to TOGAF® – and ArchiMate® – are a bit different from how **Object Management Group** (**OMG**) defines these two keywords. This usually confuses the stakeholders and results in the enterprise architects communicating using a language that no one else understands and delivering presentations that no one gets.

High cost-to-value ratio

Enterprise architects usually charge high rates and EA projects take long periods of time before anyone sees tangible deliverables, especially if done in a waterfall approach. According to the **Glassdoor** careers website, the average annual salary of an enterprise architect is about $110,000 in the United States (`https://www.glassdoor.com/Salaries/enterprise-architect-salary-SRCH_KO0,20.htm`). This number is an average and can go up to $220,000 per year for a senior director enterprise architect position according to the same source. If you have a team of four working in your EA office, that will cost your organization nearly $1 million per year if we consider the additional benefits and cost overheads. Additionally, most EA tools are very costly to procure with average license prices of around $10,000 per user (we will talk about EA tools later in this chapter).

Taking into consideration all the aforementioned points, the outcomes of EA projects can be bulky documents that are expensive, theoretical, boring to read, difficult to understand, and do not help in resolving any of the ongoing issues. The decision-makers find it difficult most of the time to justify huge EA investments for the relatively small added value.

TOGAF® is mostly adopted by IT people

Even though EA is an enterprise-level practice and aims to align business and IT within the organization, the reality is that it has mostly been adopted by IT people. This by itself is not an issue because life is full of examples of people who succeeded in different careers and people who quickly gained the required skills and expertise in new domains. IT professionals have always proven that they can lead the trending wave even if it is not just for IT – project management and EA are two examples of that.

The real issue is that IT enterprise architects kept focusing on IT operation and software development, and for them, every deliverable or artifact must serve as an IT solution. You can see this very clearly when browsing the job descriptions of enterprise architect positions on any recruitment website. You will see that most of the offered positions are titled enterprise architect, but the description is mainly looking for deep software development and programming skills or core network operations skills. This usually ends up with EA as a subdivision under the IT unit and the enterprise architect reporting to the IT manager or the **Chief Technology Officer** (**CTO**) in the best-case scenario. EA in this case will be limited to IT and will fail to achieve the most important goal it is established for, which is bridging the gaps between business and IT and aligning IT to business strategies.

After having this quick overview of TOGAF®, highlighting its benefits and drawbacks, it is time to introduce a solution that utilizes all of the benefits keeping you away from the drawbacks.

Introducing agile EA

To start with, being agile in practice does not necessarily mean that we will follow any of the agile software development methodologies, such as **Scrum** or **eXtreme Programming** (**XP**). Being agile simply means being adaptive to the continuous changes to the requirements within your enterprise and being responsive with the right amount of information, at the right time, to the right people.

Another important point to keep in mind is we are not introducing a new methodology that is better or worse than TOGAF®, and we are not following the **Open Agile Architecture**™ either (`https://pubs.opengroup.org/architecture/o-aa-standard/index.html`). We are still following TOGAF®; it will still guide our progress and we will be reusing more of the framework's components, such as the taxonomy, metamodels, TOGAF®'s techniques and guidelines, and most of the ADM content. The main differences will be in the order of the ADM phases and steps.

This section will explain how EA can provide better value if implemented with agile principles in mind by focusing on delivering tangible and valuable artifacts that can grow as the work progresses. Before we go deeper into the approach, let's start by understanding what agile EA means.

Understanding agile EA?

Part of our proposed solution is to be agile in the EA practice and to focus on delivering products where they are needed the most. These products are called **architectural artifacts** within the context of EA, and they include all the diagrams, catalogs (dictionaries), and matrices (cross tables) that make the documents that are produced during an EA practice. I highly recommend that you read the *12 agile principles* that are defined in the **Agile Manifesto** for your knowledge if you are not already familiar with them (`https://agilemanifesto.org/`) but replace every *software* word you see with *artifact* to see EA through agile eyes. The agile principles can be applied to any project in life, including personal ones, and not just on software development.

As an enterprise architect, people will expect you to provide them with advice and solutions everywhere in the enterprise. If they feel that you will slow their progress down or enforce things that they do not require, you will end up detaching yourself from the rest of the organization and will end up sitting alone in your *ivory tower*.

One main reason for EA failure in many organizations is that product owners overlook involving the EA office in their projects to avoid possible delays. Product owners want their products to be up and running as soon as possible. They will not feel comfortable adding tasks to their scope of work that they do not believe are needed just because the framework says so. They will not be happy if you ask them to hold their progress and wait until the EA office finishes tailoring the framework, defining value streams, and defining the governance model.

Defining – or even just refining – all these components can keep you and the rest of the team busy for months, if not years, and the product owner will soon perceive you as an obstacle and a risk to the project. It is a fact that you need to respect if you want to have a successful cooperative relationship with the rest of the enterprise.

Following a methodology is a way to do things right, no doubt about it. However, the product owners will not wait for you to finish tasks that they do not require and are not part of their project's scope of work. This is where you need to balance *doing things right* and *doing the right things*.

If the objective is to develop a new web application, for example, as an enterprise architect you must complete the full picture by connecting all the dots and making sure that the application realizes actual business services, and the application has – or will have – the required technology infrastructure that will support it. The key to being more practical than theoretical is setting up the scope and prioritizing your tasks to deliver useful artifacts within the available time.

Continuing with the web application example, it is more useful and more appreciated by the product owners if you start by conceptualizing the application services that they have in mind and help them to plan and design them. Once you have the desired artifacts, you can add more content to them and/or develop other artifacts that are based on them. You will keep using TOGAF® and its material as a reference for the big picture, but you do not have to start from the top, and you do not have to complete each phase before you start the next one. That is the waterfall approach, which was deprecated from software development years ago and must be deprecated from EA development as well.

Comparing agile EA with EA

The idea of agile EA is not new, but in fact, it is as old as EA itself. John Zachman, the father of EA, introduced his framework in the mid-1980s and it is still known today as the Zachman Framework. Every EA practitioner or learner will have heard of it even though the number of Zachman Framework practitioners is far fewer than those of TOGAF®. The Zachman Framework is incredibly famous for its 6x6 matrix, which has *What*, *How*, *Where*, *Who*, *When*, and *Why* as columns, and has enterprise layers such as *Scope Contexts*, *Business Concepts*, *System Logic*, *Technology Physics*, *Tool Components*, and *Operational Classes* as rows (as defined in `https://www.zachman.com/about-the-zachman-framework`).

Unlike TOGAF®, the Zachman Framework has no specific process to follow, and architects are free to start anywhere they want on the matrix. Once they define the artifacts they need, they can simply go to any other cell on the matrix and build more artifacts. This is an agile approach that was introduced way before agile software development was even thought of. Architects can start with the process definition, for example, followed by responsibility configuration and then the inventory instantiation artifact without constraint. They can decide which artifacts to develop based on stakeholders' demands and requirements and the availability of information, rather than deciding based on a sequential flow of steps.

The big picture will be formed as you build more artifacts here and there and as more details and relationships are revealed, exactly like finding and fitting the pieces of a puzzle. Sometimes you will find that you have made some wrong assumptions based on the information that you had in hand at that time, and the artifact is incomplete or unclear, but this is a very acceptable side effect of agile development, and you need to keep in mind that nothing is written in stone and every artifact is subject to changes.

Sharing your artifacts continuously with the rest of the team and with the product owners as you gradually build them will help in communicating your thoughts in the early stages and will help you make the required corrections and adjustments as needed. Making continuous small changes and updates to the artifacts is better than going through a lengthy waterfall process and getting caught in *Effort Blackholes*.

Embedding agile EA into TOGAF®

As mentioned earlier, this book is based on TOGAF®, and the reason we talked briefly about Zachman in the previous section was to show that the agile way of EA development is not a strange idea or an extreme thought to be afraid of. In this section, we will show you how you can embed the agile mindset into TOGAF® without compromising the principles of the framework. These are the main elements of the agile EA approach:

- The big picture will remain guided by TOGAF®.
- Start anywhere where EA contribution is required.
- Focus on delivering artifacts.
- Use smaller metamodels.
- Build the architecture governance as the architecture evolves.
- Use an EA tool for the architecture repository.

The following subsections will explain each of the preceding elements in more detail.

The big picture will remain guided by TOGAF®

Being agile and focusing on artifacts does not mean losing sight of the big picture. Everything that is in TOGAF® will still be used as guidance and this is how:

- TOGAF®'s content metamodel will still be the big picture that shows the different elements within the enterprise and the possible relations between them. We will use ArchiMate®'s metamodel because its metamodels provide more details than the ones provided by TOGAF®, but both framework models have the same background.

- We will still use the taxonomy of TOGAF® and ArchiMate® as the formal definition of the enterprise elements. We may elaborate on some definitions if needed, but we will stick to the standard definitions all the time.

- We will still use the list of artifacts and the list of proposed inputs and outputs by TOGAF® as guidelines.

- We will still use TOGAF® for guiding the definition of our governance model but without forcing it. This means that not every role, process, or responsibility must be built upfront, but we take what we need and change as we progress.

- We will adapt and reuse any of the tools and techniques that are provided by TOGAF® and adjust them to our needs as we progress.

Start anywhere where EA contribution is required

As an enterprise architect, you need to contribute to every development progress for any business unit and at any layer of the enterprise. Whether it is a strategy-crafting initiative, a marketing campaign, a new software procurement, or a plan to retire some legacy mainframe applications, enterprise architects need to be involved in all these initiatives as they progress. They do not have to be the subject matter experts in all these different domains, but they need to add value to all these areas by encouraging the use of standards and connecting and aligning the different components in the different layers. Involving the enterprise architects in multiple projects at different locations of the enterprise can uncover some hidden relationships between these projects that only someone with EA eyes can see.

To demonstrate how enterprise architects can contribute to any layer with great value, let's take the mainframe application retirement as an example:

- Retiring an application will affect the other applications that exchange data by taking the forms of inputs, outputs, or both. If not taken into consideration, this integration dependency may get broken and will result in unplanned errors or missing information.

- It will affect the infrastructure (the technology) that the application uses so either it becomes available for other applications, or it becomes useless and ready to be retired.

- It will affect the people operating, administrating, and using the application so either they will require training to be able to use the replacement application or they need to be reallocated and play different roles for the best use of resources.

- It will affect the business services that depend on the application in automating some or all of the processes. If multiple business services depend on the application, there will be a possibility of some of them going down or being interrupted due to the broken dependency.

- Having a service or services down may affect the organization's ability to achieve its strategic goals, which may by itself have effects on other elements in other layers because everything within the enterprise is connected.

Only enterprise architects can provide these dependency views, and this is one example of where they can add value to any project at any layer. We will talk more about this in detail in *Chapter 8*, *Business Architecture Models*.

Focus on delivering artifacts

One benefit of following an EA framework is to make deliverables and artifacts standardized, organized, consistent, and of higher quality. If the delivered artifacts are of no use to the stakeholders, then your efforts and their time will be wasted, something that EA has struggled with for years. Remember that any effort that does not have a deliverable is a wasted effort. There is the concept of delivering the **Most Viable Product** (**MVP**) in agile software development, having the minimum essential features delivered first, then the product keeps growing by adding more features to it. That concept must be followed during EA development as well and the product within the EA context is an artifact.

One of the recommendations in agile software development is to break down the tasks into smaller pieces or parts, and we will follow this in agile EA. Having a deliverable such as the **Architecture Definition Document** (**ADD**) that contains multiple smaller artifacts, such as the business processes catalog, business services catalog, application components catalog, business processes to application components matrix, application components to technology services matrix, and many other sections, it can take an exceedingly long time with all the different sections that it contains. So, instead of having a single task for developing the ADD, it is better to break it down into smaller artifacts that each can be completed within a few days or 2 weeks at most, or else they must be broken down into smaller parts that fit within the 2-week window.

If you have multiple engagements to participate in, then you need to distribute your time spent on all these engagements without affecting or slowing down the others. You can keep growing your artifacts simultaneously as you go, and you may discover some dependencies between the two that no one considered before.

Use smaller metamodels

Metamodels are the templates that architects use as references when creating EA artifacts to maintain the consistency and the quality of the artifacts across the EA repository. TOGAF® and ArchiMate® have generic metamodels that you can start with, but they do not cover every possible relationship and sometimes they are difficult to read and understand. Customizing the standard metamodels is recommended to fit the exact requirements of your organization; however, trying to tailor the entire TOGAF® and ArchiMate® metamodels can take a very long time. Following the same concept of breaking large artifacts down into smaller ones, you must break the effort of tailoring the metamodel down into smaller parts.

The next section of this chapter will introduce the concept of *focused metamodels*, which we will be following throughout the rest of the book, but for now, remember that creating metamodels is as important as creating the artifacts themselves. Even if you are the only architect who is building artifacts, having metamodels will help you maintain the required consistency within your own set of deliverables.

Build the architecture governance as the architecture evolves

Defining the architecture governance model is essential to ensure the proper development and maintenance of EA artifacts. With proper governance, every person involved in the architecture development and maintenance knows what to do to keep the artifacts updated. Being agile does not mean being chaotic and without processes or strategy. Building an architecture repository without a governance model around it will result in having most of the artifacts outdated or with broken relationships within a year at most. Outdated or incorrect information will break the trust between stakeholders and the information that the EA repository provides, so it must be avoided. Therefore, we know the importance of having an established architecture governance model but as with everything else, it must be broken down into smaller parts and we start with only the ones that we need for the MVP, then we enhance it by adding more to it.

Chapter 37 of TOGAF® 9.2 (`https://pubs.opengroup.org/architecture/togaf9-doc/arch/chap37.html`) provides the concept of the EA repository, and *Chapter 44* (`https://pubs.opengroup.org/architecture/togaf9-doc/arch/chap44.html`) provides the concept of architecture governance. Both chapters represent the visionary big pictures that you can aim for. However, that concept can be too big and difficult to implement if you are just starting, so you need to build the governance MVP that can govern only what is being developed and avoid *boiling the ocean*. Attempting to boil the ocean will result in *Effort Blackholes* with no doubt. It is important to look at the big picture and know what is behind it, but it is more important to use your resources wisely. *Chapter 10, Operating the EA Repository*, of this book will provide you with a simple governance framework that can evolve as per the maturity of your organization.

Use an EA tool for the architecture repository

It is very possible to implement TOGAF® or any EA framework using Microsoft Office tools and hosting everything on a shared network drive. However, having an EA tool such as Sparx makes it easier and more practical to do so, especially if you are doing it in an agile way. Since agile is all about accepting changes even at later stages of development, doing that without a tool can be extremely difficult to maintain. The tool will let you know if you are changing an element that is connected to other elements, for example, which may break the established relationship. We will talk more about the EA tool that we will use to build the EA repository in the last section of this chapter. For now, you need to keep in mind that having an EA tool makes modifying the artifacts easier, which will help you do EA in an agile way.

Now you know the agile EA approach and the importance of building artifacts, it is time to introduce the modeling language that will be used for this purpose.

Introducing ArchiMate®

The ArchiMate® is a visual modeling language that is published by The Open Group® especially to be able to model and create EA artifacts. It is a complementary extension to TOGAF®, so it provides architects with an extended metamodel, extended taxonomy, and a visual modeling notation.

> **Important Note**
>
> It is highly advisable to get yourself familiar with ArchiMate® if you have not used it or read about it before because all of the information you need for that is already available online. The ArchiMate® 3.1 Specification is easy to navigate, online, and free to all content that is published by The Open Group®, the same publisher of TOGAF®: `https://pubs.opengroup.org/ architecture/archimate3-doc/`.

Architects are free, of course, to use any modeling language they are familiar with to build their artifacts, such as the **Unified Modeling Language (UML)**, **System Modeling Language (SysML)**, **Business Process Modeling Notation (BPMN)**, **Data Flow Diagrams (DFDs)**, and **Entity Relationship Diagrams (ERD)**.

Each of these modeling languages is good for a specific set of artifacts but you will start to struggle when trying to mix elements from different architecture layers within the same view, which is what EA is all about.

Let's see why ArchiMate® is preferable over the other modeling languages for developing EA artifacts.

ArchiMate®'s role in EA artifacts

If you are modeling for the purpose of software development only, then UML and SysML would be ideal, and people have used them for years. However, if you are developing EA artifacts, you may need to model strategy and motivation elements along with the software components, which can be a bit difficult to do without some workarounds. The same thing is true when you chose BPMN for modeling business processes. BPMN is a familiar notation for most business and IT stakeholders, which makes it the default choice for modeling business processes. However, you cannot use it to model software or hardware components, or to model data structures and object relationships.

This is where ArchiMate® comes to the scene and is gradually becoming the choice for many enterprise architects. With ArchiMate®, you can have elements from the Technology Architecture layer (a node, for example) serving an element from the Application Architecture layer (an application component, for example), which serves a component from the Business Architecture layer (a business service, for example), which realizes an element from the Motivation Layer (a goal, for example), all in one diagram. There is no other modeling language than ArchiMate® that can handle this mixture in one place:

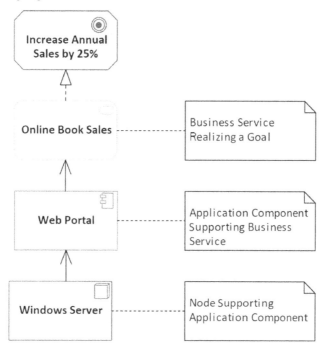

Figure 1.2 – An ArchiMate® example showing elements from different layers

To be fair, with UML, you can extend the basic UML elements such as the classes and the components using stereotypes to get any EA element you need. In the preceding example, you could have used a *class* with the <<goal>> stereotype, a *class* with a <<business service>> stereotype, a *component* with the <<application component>> stereotype, and a *component* with the <<technology component>> stereotype, respectively.

You must, however, limit the list of stereotypes that you will use to the elements that you have defined in your metamodel to avoid having too many stereotypes and losing the meaning of elements' identification and classification. Additionally, you need to make sure that you consistently follow the same spelling for stereotypes. You are free to use <<application component>>, <<ApplicationComponent>>, <<Application Component>>, or any other way you prefer, but you must stick to the same within the entire EA repository or else you will end up having multiple different classes representing the same type of elements. This is where metamodels become very handy because you can always refer to them to check which format you used. The following diagram shows how you can model the same information in the previous ArchiMate® diagram using UML and stereotypes:

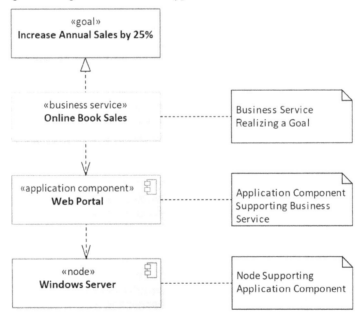

Figure 1.3 – An example of EA artifacts modeled using UML

Notice that UML does not have the *support* relationship, so we must replace it with the *dependency* relationship and change its direction. The preceding diagram can now be read as the goal is realized by a business service, which depends on an application component, which depends on a node (or a technology component).

> **Tip**
> Always keep UML as a possible option in your mind for modeling EA artifacts as many organizations prefer to use it instead of introducing a new modeling language such as ArchiMate®. A new modeling language will require additional change management efforts that will be translated to additional investment in time and money.

The good news is that everything you will learn from this book can be done with UML. I have built more EA repositories using UML than with ArchiMate®, so it is a quite common and acceptable approach.

Understanding ArchiMate® modeling specification

The ArchiMate® specification has been defined in multiple versions by The Open Group® but the latest is version 3.1 and it is the one that we will be following. We will not repeat the information that is available online, but we will use some of the elements in more detail and we will map them to TOGAF®'s artifacts. In this section, we will cover the following:

- ArchiMate® 3.1 element hierarchy
- Elements color theme
- ArchiMate® modeling notation

Let's explore in more detail each one of them.

Exploring ArchiMate® 3.1 element hierarchy

ArchiMate® categorizes elements into two main categories regardless of the layer: structural and behavioral. **Structural elements** are the elements describing objects and entities; therefore, they are also known as nouns. Structural elements can be subdivided into active and passive. **Active structural elements** are the ones that perform the actions while **passive structural elements** are what actions are performed on. This makes application components, for example, active structural elements while data objects are passive structural.

Active structural elements can be further subdivided into internal and external elements based on how exposed they are to the surrounding environment. An application component can be accessed only through an application interface, for example. Whether the interface is a **User Interface** (**UI**) or an **Application Programming Interface** (**API**), actors must use interfaces to communicate with components.

Behavioral elements describe the actions that structural elements can perform, so you can think of them as the verbs that nouns can perform on other nouns. Behavioral elements can also be subdivided into internal and external elements. **Internal behaviors**, such as processes, describe how the structural elements get the job done. **External behaviors**, such as services, describe to the external environment what the structural elements provide to them.

Confused? Do not worry because this part was meant to help you in reading the official ArchiMate® metamodels, which we will replace with a much easier-to-read metamodel, so please be patient. Just explore the following diagram, which summarizes the hierarchy of ArchiMate® 3.1 elements, but you do not have to remember any of it for now:

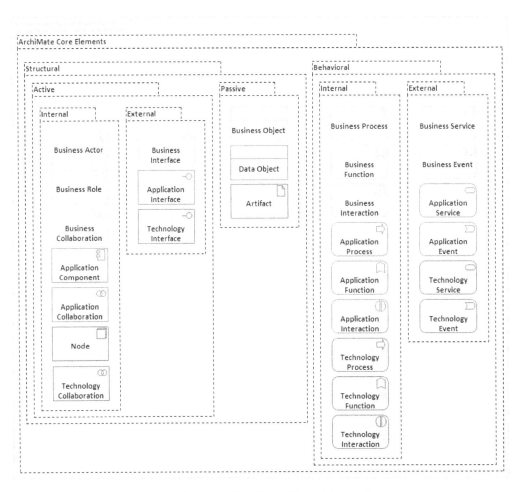

Figure 1.4 – The hierarchy of ArchiMate® 3.1 core elements

The layers and the elements in the preceding diagram are known as the **core elements**. There are additional layers and elements, but our focus will be on the core elements, which we will use most of the time. We will explore other non-core elements as we need them.

Exploring elements color theme

There is no strict color theme that ArchiMate® forces architects to follow, so you are free to follow any coloring theme or simply make everything in black and white as long you consistently follow the same practice in all your diagrams. If you chose blue, for example, for application layer elements, then it's a good idea to stick to the exact same shade of blue in all the diagrams throughout the entire repository. Using the same color for elements that belong to different architecture layers can be confusing. You can also use black and white for everything to avoid complicating the rules for you and/or your team.

Also keep in mind that some regulations may require using specific color codes and some regulations restrict the usage of colors to differentiate between objects, so make sure that you comply with the regulations of the region you live in.

As a result, it is better to use the named ready-made colors instead of creating custom colors that require remembering their RGB codes every time you use them. Your metamodels can also tell the colors of the elements that need to be followed across the repository.

> Note
>
> If you are reading a printed copy of this book, then some diagrams will be in grayscale. For **full-color versions** of these diagrams, please refer to the graphic bundle available here: `https://static.packt-cdn.com/downloads/9781801076166_ColorImages.pdf`.

ArchiMate®'s online documentation consistently uses *yellow* for business, *blue* for application, and *green* for technology layer elements. The entire shapes are filled with the proper colors surrounded by black borders:

Figure 1.5 – The color theme of ArchiMate®'s online content

We will follow a similar coloring theme in this book, but we will fill the elements with white and apply the colors to the borders. We will also use orange for business elements instead of yellow for better visibility:

Figure 1.6 – The color theme of this book

The color names of the preceding are defined in Sparx as follows:

- **Orange** (RBG 255, 165, 0); will be used for all Business Architecture elements

- **Royal Blue** (RBG 65, 105, 225); will be used for all Application Architecture elements

- **Dark Green** (RGB 0, 100, 0); will be used for all Technology Architecture elements

As mentioned earlier, it is up to you to decide what colors to use as long as you and other EA team members remain consistent with your choice.

Introducing ArchiMate®'s modeling notations

ArchiMate® provides two **notations** to represent each type of element. As you can see in *Figure 1.7*, one uses rectangular shapes with small icons in the top-right corner of the rectangle (**Rectangular Notation**), and the other uses large versions of the icons without the rectangular borders (**Borderless Notation**). ArchiMate® uses shapes rather than textual stereotypes to visually differentiate elements, which can be confusing at the beginning until you get used to all the different shapes.

Have a look at the following diagram and notice the differences between the two notation styles. The top row uses the rectangular notation to model an application interface, an application component, and an application function, respectively. The small icons in the top-right corner of each rectangle represent the type of element. The bottom row uses the borderless notation to model the same three elements we mentioned previously in the same order.

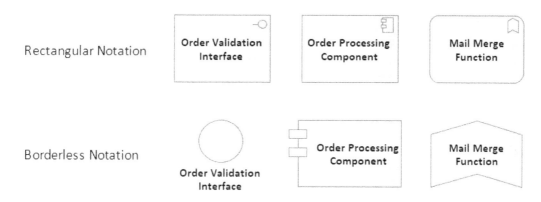

Figure 1.7 – ArchiMate®'s two notations

There is no difference at all between the two notations; it is just a matter of preference. You can switch between the two styles in Sparx by simply right-clicking on the shape and checking or unchecking **Advanced** > **Use Rectangle Notation** from the context menu. The captions will always be inside the rectangular shapes, while they may get placed outside the borderless shapes for some elements. If you are modeling a **composition** relationship, you can do this by placing one element inside the other, and the container element will be the **whole** or the **parent** and the contained element will be the **part** or the **child**. In the composition scenario, I recommend using the rectangular notation for the container to have more space inside.

In general, avoid using the two notation styles in the same diagram for the same element type as this will confuse your audience. For example, do not use the rectangular notation to represent some application interfaces and use the borderless notation to represent some other application interfaces on the same diagram because this will indicate that there is a difference between the two. We will explore the two different notations as we go, and we will use one style in some diagrams and use the other style in some other diagrams but will never mix the two in a single diagram.

Metamodels

ArchiMate®'s has extended TOGAF®'s metamodel by adding more elements and more relationships. But unlike TOGAF®'s metamodel, where all the elements and relationships are displayed in a single diagram, ArchiMate®'s metamodel is divided down into smaller but closely related metamodels, so there is a metamodel for the motivation layer, a metamodel for the business architecture layer, a metamodel for the application architecture layer, and so on. The reason – I believe – is to avoid having a big, busy diagram that is difficult to understand.

> **Important Note**
>
> Please refer to *Figure 30-5* from TOGAF® 9.2 for the core content metamodel, and *Figure 30-7* for the full content metamodel (`https://pubs.opengroup.org/architecture/togaf9-doc/arch/chap30.html`). You need to make yourself familiar with how TOGAF® and ArchiMate®'s metamodels are defined and how can you interpret them as this is an essential step for applying the standards to ensure that the architecture content is in full alignment with the framework.

In this section, we will talk about the following:

- ArchiMate® 3.1 metamodels
- ArchiMate® taxonomy

The following subsections provide more details for each.

ArchiMate® 3.1 metamodels

ArchiMate® metamodels are written in a very abstracted way in order to provide wider coverage of enterprise elements in a smaller space. They contain elements such as the **Application Internal Behavior Element** to represent all descendant elements, rather than putting each of them on the diagram and complicating it with intersecting relationships. Check this ArchiMate® metamodel online to be able to follow my explanation: `https://pubs.opengroup.org/architecture/archimate3-doc/chap09.html#_Toc10045390`.

As you can see, the metamodel tells you a lot of things that can help you when you are modeling application layer components, and they have tried to make it as compact as possible. If you want to model an application component, for example, the metamodel will tell you what other elements can connect to or be in a relationship with your component.

As you can see, there is no application component in the diagram but the more generic class, the **Application Internal Active Structure Element** class, which can be confusing for those who are new to ArchiMate®. It tells us that application components can be assigned to an application internal behavior element, which can be translated to *the application component that can execute the application process.*

It will take you some time to be able to quickly interpret the ArchiMate® metamodels, but the good news is that we will be presenting metamodels differently and easily throughout this book.

The ArchiMate® taxonomy

Since ArchiMate® and TOGAF® are published by the same organization, you're better off thinking of ArchiMate®'s taxonomy as an extension of TOGAF®'s taxonomy rather than two different ones. The definitions of the two are very similar and in many cases, they are exactly the same. But because TOGAF® is more generic and ArchiMate® is more detailed, you may find that a single definition in TOGAF® has multiple specific definitions based on the architecture layer they belong to. For example, TOGAF® has a single definition for **process**, while ArchiMate® has three layer-specific definitions for it: **business process**, **application process**, and **technology process**.

We will include both definitions when introducing new elements and will try to add more elaborate descriptions supported with examples to clarify them. Some definitions are written in a difficult-to-understand language, so we will try to simplify that. Definitions will be quoted as is from the sources and will always be in *italics* to be easily differentiated from the rest of the text. The following is an example of the definition of a business process:

A **business process** *represents a sequence of business behaviors that achieves a specific result such as a defined set of products or business services* (ArchiMate® 3.1: `https://pubs.opengroup.org/architecture/archimate3-doc/chap08.html#_Toc10045374`).

Any additional elaboration and supportive examples will usually come in the paragraph following the definition, like this paragraph.

The next section will introduce Sparx Systems Enterprise Architect as our tool of choice for modeling EA artifacts.

Introducing our focused metamodels

Reading about and understanding the ArchiMate® metamodels can make the difference between a true architectural artifact and any other diagram. But as you saw, it is not a straightforward step to get the information you want. I used to use paper and pen to decipher the metamodels into something that I can use in my models. When it comes to connecting to elements from other layers, you must refer to the metamodels that define the cross-layer relationships to get the bigger picture.

Since this book is all about making things more practical, we are providing you with metamodels that are focused on one element at a time. The following diagram represents the application component-focused metamodel and I bet it looks way easier to read and understand than the standard ArchiMate® metamodel:

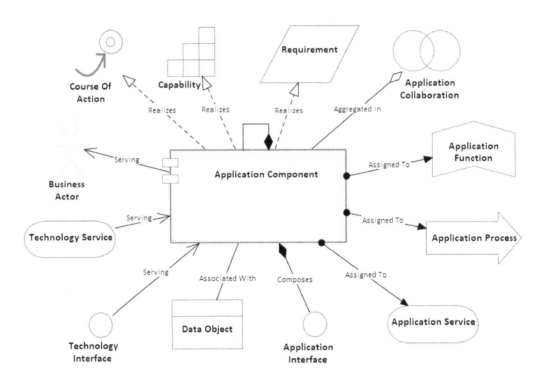

Figure 1.8 – Application component-focused metamodel

You can immediately tell that an application component can be assigned to an application function, application process, and application service. It can compose application interfaces and can compose other application components as well. We will cover in more detail all the components that are included in the focused metamodel.

Following a metamodel is essential in unifying the EA artifacts that are developed by different architects. They are the **blueprints** for developing models, and without them, each diagram will look different and be inconsistent. You do not have to include all the elements that are shown in the focused metamodels in every model you develop. It is a reference that shows every relationship possibility, so you take only the ones you need.

Additionally, do not forget that we are following an agile way of implementing EA, so we do not have to develop all the focused metamodels before we start modeling as this is a waterfall approach. As we start following the example scenarios in *Section 2* and *Section 3* of this book, we will build only the metamodels that we will need. Even if we do not cover every element in the ArchiMate® specification, you should be able to build the metamodels for the remaining ones and populate your own EA repository in the same way that we will be practicing.

Introducing Sparx Systems Enterprise Architect

Sparx Systems Enterprise Architect – known as Sparx – is a modeling tool that has evolved over the years from just UML modeling in the early 2000s to a tool that can today model almost everything, including modern AWS and Azure diagrams. Note that this book is not sponsored by Sparx Systems and not affiliated with any of their partners; we have chosen Sparx for multiple reasons that will be explained in the next subsection, and we will have a first look at the components of its UI in the subsection after that.

Why Sparx?

Using an EA tool is not mandatory to establish an EA office or build an architecture repository, but it makes things easier, more efficient, and better governed. If you or your organization are not using an EA tool, I highly advise you to do so unless you are still in the early discovery phases.

If you have not used Sparx before, you can download a trial copy now and get familiar with the different components of the UI, as well as getting used to the way of creating diagrams. Sparx Systems has a huge library of resources that can help you learn their product and become an expert in it. It will be extremely helpful to explore the fundamental materials provided at this location before you proceed with the book (`https://sparxsystems.com/resources/user-guides/15.2/index.html`).

If you use a different tool other than Sparx, you will still be able to apply all the concepts that will be provided in this book in your tool of choice. The concepts can be applied using any tool, but we will be focusing on Sparx only. The reasons that we have chosen Sparx can be summarized as follows:

- Low cost of ownership
- Direct download and free trial
- Widely known

Let's explore each one of these advantages in more detail.

Low cost of ownership

Sparx is a tool that anyone can purchase without massively hurting their pockets. If you are studying and practicing at your own cost, you can get Sparx up and running for as little as $229. Most other EA tools cost an average of $10,000 per user per license, so you can see how huge the difference is. If you work for an organization that does not want to put a big investment in an EA tool, either due to budget constraints or to lower the risk involved with practicing EA, then Sparx is an excellent tool to start with. Once the *architecture maturity* of your organization increases, you can choose to migrate to a more sophisticated and more expensive tool if needed.

Please check the Sparx Systems website (`https://sparxsystems.com/`) for an up-to-date price list and to learn more about the different licenses they offer. For the record, I am using the *corporate* license for this book, so every diagram you will see in this book – and many more – can be done with it.

Direct download and free trial

If you want to explore Sparx, you do not have to go through a long procurement process where you fill out a form with your information, submit it, and wait for the marketing team to call you back, ask you more questions, schedule a demo, and make a purchase order before putting your hands on their tool. With Sparx, all you need is to register your email address and you will be provided with a link to download a fully functional, 30-day trial copy of the EA tool. If you do not like it, no one will keep bugging you with sales emails and phone calls; just simply uninstall it. If you like it, pay for the license online, enter the provided license key, and your trial version will become a licensed version. Everything that you have practiced so far will still be there for you.

Widely known

Because of the previous reasons, Sparx is a very well-known tool in the market and almost every architect has used it or heard of it. Throughout my career of EA implementation, Sparx has been the tool that more than 40% of my clients have used, about 10% used other tools, and the rest were not using EA tools at all. You can do an internet search for EA tools and see what is currently available on the market.

Having talked about why we're using Sparx, let's explore the main areas of its UI and get ourselves familiar with it.

The Sparx UI

This section will give you a very quick overview of the Sparx UI. The following figure shows you the sections of the Sparx UI followed by a brief description of each. We have used version 15.2 of Sparx in this book, so if you are using a different version, you may see some differences between what is in this book and what is on your screen.

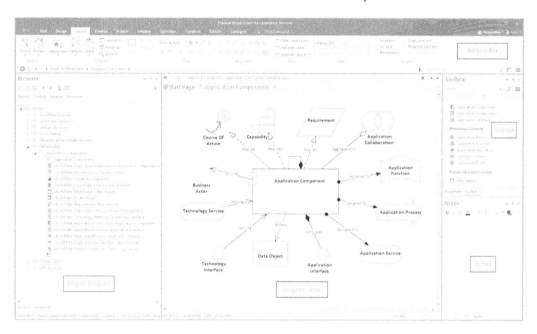

Figure 1.9 – The Sparx UI showing the main sections

As you can see in the figure, there are four main areas of the Sparx UI:

- The **Diagram Area**
- The **Project Browser**
- The **Ribbon Bar**
- The **Toolbox**

Note that the project browser area has multiple tabs (currently showing the **Browser** and **Inspector** tabs), and the **Toolbox** area has multiple tabs too (currently showing **Toolbox** and **Properties**). These tabs can be docked to any side of the screen, so you can drag the **Toolbox** tab, for example, to fit within the same area that the project browser fits. In fact, you can fit all the tabs on one side of the screen and have more space for diagrams in the middle. It is all up to your preferences.

Let's explore each of these main areas now.

The Diagram Area

The diagram area is where your diagrams get created and designed. There is nothing much to say about this area right now, but this is the area where the actual work is done, so we will learn more about it as we progress. Right-clicking on the diagram background will bring a context pop-up menu. Click **Properties** and the diagram properties window will be displayed:

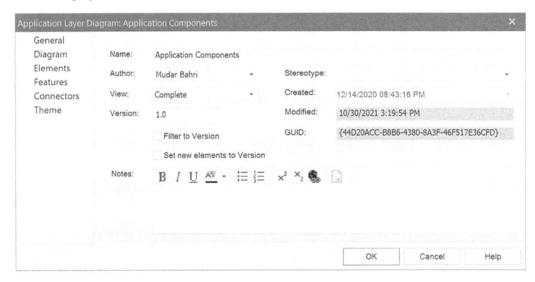

Figure 1.10 – The diagram properties window

Please make yourself familiar with the different tabs and options and refer to the online material whenever needed by clicking on the **Help** command button or pressing *F1* on the keyboard. You do not have to remember every screen option now, but just explore what is there.

Project Browser

This area displays your resources in the form of **packages** and **elements**. Think of the Windows filesystem, for example; the packages are like folders and elements are like files. In a nutshell, everything in Sparx is either an element or a package. Diagrams are elements, and all the shapes that you place on them are elements too.

Unlike in visual modeling tools such as Microsoft Visio and Lucidchart, in EA modeling tools including Sparx, you create an actual element for every shape you put on the diagram, so the shape is not only a visual item but an actual element that lives inside your repository. Take a second look at *Figure 1.9* and you will notice that for every element that we have placed on the diagram, there is a representation of the same element in the project browser, including an element for the diagram itself. This is an essential concept in EA modeling because if you want to show multiple views (that is, diagrams) for the same element, then you have to reuse the element from the project browser, rather than creating a new element from the toolbox. If you find this confusing, do not worry now as we will practice many examples that show you when to *create* an element from the Toolbox and when to *reuse* it from the project browser.

Explore the properties window for any elements by highlighting the element and right-clicking it, then choosing **Properties** from the context menu. You will notice that it is the same window for all elements regardless of their type:

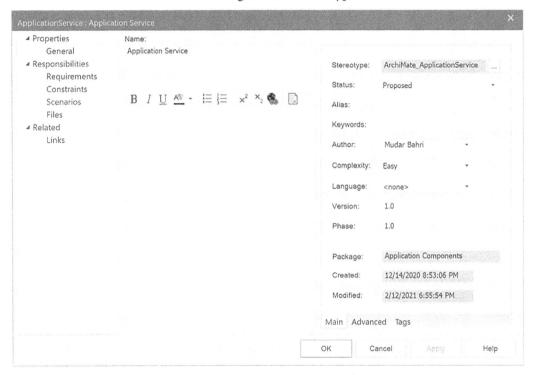

Figure 1.11 – Element properties window

Explore the different tabs and options and make yourself familiar with them. Refer to the online documentation if you need help.

The Ribbon Bar

Sparx has Microsoft Office-like toolbars with large icons and menu options. Some ribbons are for configuring the Sparx environment, some are for alignments and layouts, some are for publishing and exporting, and other options. The following figure shows four sample menus, but you need to refer to the Sparx online user guides for the complete documentation of all the options and actions they provide. We will only talk about the ones that we will use, so not every menu option will be covered:

Figure 1.12 – Four sample toolbars from the Sparx ribbon bar

Notice how some menu items have a small arrow indicating that clicking on them will open a drop-down list of options. Also notice that menu items that perform similar functions are grouped within groups. The **Start** toolbar, for example, has four groups: **Explore**, **Desktop**, **Collaborate**, and **Help**. Throughout the book, we will follow this pattern to guide you on which menu item to select: **Toolbar Name** > **Group** > **Menu Item** > **Menu Sub-Item**, which is the same style that Sparx uses in the online material. For example, if you need to adjust the Z-order of an element and bring it to the top, I will ask you to click **Layout** > **Alignment** > **Bring to Top**, which makes it easier for you to find that option than just saying "Click on the **Bring to Top** menu item."

Toolbox

The toolbox has dynamic content, and it changes according to the type of the diagram. The toolbox that you see in *Figure 1.9* is the one associated with diagrams of the ArchiMate® application diagram type. You can change to different content by clicking on the hamburger menu (the menu that has a stack of three horizontal lines) on the top-right side of the toolbox and changing to different content, as shown in *Figure 1.13*:

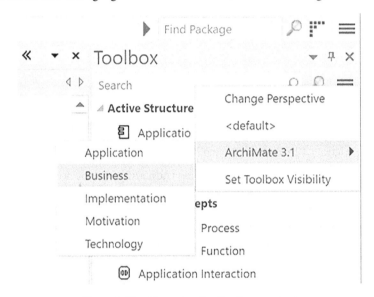

Figure 1.13 – Changing the Toolbox content

As we have advised for other areas on the UI, if you need to explore the **Toolbox** area, use **Help** if needed and get yourself familiar with it.

> **Important Note**
> Drag items from the toolbox to the diagram area to create new items in the repository. Drag items from the Project Explorer to the diagram area to reuse an existing element.

This was a lot of information to put in one chapter, but more details will come as we practice putting this information into useful artifacts.

Summary

In this chapter, we have laid the foundations of the structure of the book and introduced the core concepts that we will address in the remaining chapters. This will have helped you to understand the benefits of implementing an industry-standard framework such as TOGAF® but without forgetting that nothing comes without problems and that things can go wrong as well. You should have realized the importance of bringing the agile mindset to the EA world without losing the big picture. We have also chosen ArchiMate® as the modeling language and Sparx Enterprise Architect as the modeling tool that we will use to build an EA repository.

In the next chapter, we will introduce three fictional scenarios to get hands-on with realistic problems that you as an enterprise architect will face or may have already faced.

2
Introducing the Practice Scenarios

In this chapter, we will introduce the three scenarios that make this book practical and related to real-life examples. The company that we will talk about is a fictional company, which we will name **ABC Trading**, but it can represent any company that you have or will be working for.

The purpose of the scenarios is to make up simple and generic stories that can fit within your real-life work environment, even if your organization is working in a different business domain. The scenarios will be detailed enough and as close as possible to problems that you have faced or will face as an enterprise architect. The goal here is to introduce our proposed agile **Enterprise Architecture** (**EA**) approach for implementing EA, not to challenge your imagination with stories that you will never deal with

The three scenarios will address three different aspects of the enterprise that concern different stakeholders, and each requires a different set of EA artifacts:

- *First scenario:* Describes an **application architecture** project where your contribution is required to help the **solutions architect** in designing and documenting a new mobile application

- *Second scenario:* Describes a **technology architecture** project where the **Chief Technology Officer** (**CTO**) needs to reduce the IT costs and needs your help in articulating some decision-supporting information

- *Third scenario:* Describes a **business architecture** project where the **Chief Executive Officer** (**CEO**) needs your contribution to a plan for introducing a new service to customers

Remember that we are applying the concept of agile EA that was introduced in *Chapter 1, Enterprise Architecture and Its Practicality*, and are not strictly following any agile software development methodology. In any Agile approach, you do not have to wait until you fully complete a milestone before starting another, but tasks get added as you progress and as the work evolves. You keep revisiting your artifacts and update them accordingly as you know more about them.

Before moving ahead into the details, let's first learn more about ABC Trading and how the book will be structured around the scenarios described.

Structuring the book around the scenarios

In this section, you will see how the scenarios are addressed in this book. The book's chapters are as independent as possible to help you go directly to where you need guidance. If you are dealing with a technology layer problem in real life, for example, and need some guidance, you can jump directly to *Chapter 6, Modeling in the Technology Layer*, and *Chapter 7, Enterprise-Level Technology Architecture Models*. If your biggest concern is using Sparx by a team of architects, then you can go directly to *Chapter 10, Operating the EA Repository*, and so on. It is all up to your preferences, but *Chapter 3, Kick-Starting Your Enterprise Architecture Repository*, will have more step-by-step instructions than any other chapter because it will be the first to introduce working with Sparx and most of these instructions will not be repeated in every other chapter. Therefore, I highly advise you to read the book entirely to get the most benefit, or to start from *Chapter 3, Kick-Starting Your Enterprise Architecture Repository*, if you prefer to skip some chapters. Keep in mind that we are imagining that these three scenarios are happening in parallel at the same time, not in the order they are written.

Let's first introduce the ABC Trading company and then we will explain how the book is structured around these scenarios.

A brief on ABC Trading

Today, all companies have business, application, and technology requirements and there are always projects here and there to address these requirements. You have been hired by ABC Trading to establish an **Enterprise Architecture Office** (**EAO**). Your goals are to establish the foundations of the EAO, create tangible value, and gain the trust of the enterprise stakeholders. You need to make the EAO a trustworthy consulting reference for all types of projects within the organization.

ABC Trading is a mid-sized wholesale network that has been in business for 50 years. They distribute their goods through a network of retail partners, and they do not have their own retail stores. However, they do have showrooms in each warehouse to showcase their goods and provide samples to retailers. ABC Trading delivers goods from their warehouses to retailers' warehouses through a fleet of trucks owned by the company.

ABC Trading has about 1,000 employees working at different business units. The IT function at ABC Trading is just enough to get the job done, but they do not have fancy up-to-date solutions and technologies. They have a mixture of a few remaining old mainframe, client-server, and web-based applications, all hosted in a centralized data center owned and managed by the company.

There was a previous attempt to establish an EAO a few years ago, but the management decided to stop the project due to low return on investment. Sparx was purchased long ago but no one really used it or got tangible value out of it. There are several licenses available and one of them has been given to you. There will be more architects to join the EAO once established but unfortunately, you will start by yourself. You will be reporting to the **Chief Information Officer** (**CIO**) who has direct contact with all other C-level executives.

The three projects that were introduced in the chapter's introduction take place at three different business units and every **Product Owner** wants to get the job done as soon as possible so they can move on to their next assignments. They are not willing to compromise the quality of work for sure, but you must keep in mind that they have no time to listen to EA speeches or to any topic they do not have an interest in. They might have already done that a few years ago when the EAO was first established and all that they remember are presentations they never understood and terminologies they never heard before. Expect their first question to you to be *what's in it for me?*, which is also known by the short abbreviation **WIIFM**. The situation is challenging, but this is your chance to show what a practical enterprise architect can do.

The structure of the book

Starting in this chapter, we will go through the three project scenarios together and build an EA artifacts backlog containing the user stories and tasks, and these user stories will be implemented in *Section 2, Building the Enterprise Architecture Repository*, and *Section 3, Managing the Repository*. For each of the scenarios, we will do the following:

1. Provide the background and introduce the problem.
2. Introduce the stakeholders of each.
3. Convert the requirements into agile user stories.

4. Identify the tasks for each user story at a high level.

5. Add the story to the artifacts backlog.

You can use any software to manage the backlog and the user stories, such as **Jira**, **Microsoft Planner**, or even Sparx itself, but for the sake of simplicity and keeping the book focused on EA and not on project management, we will use a simple bulleted list.

Here is how the scenarios will be addressed in the remaining chapters of this book:

- *Chapter 3, Kick-Starting Your Enterprise Architecture Repository*, will guide you in step-by-step instructions to develop your first artifact in Sparx Enterprise Architect. The artifact will be based on a problem in the first scenario, which is an application architecture scenario.

- *Chapter 4, Maintaining Quality and Consistency in the Repository*, will teach you how you can create a reference model that you can use to develop diagrams similar to the first one you created. This reference model will be the first in a series of similar models, and we're calling them the **focused metamodels**.

- *Chapter 5, Advanced Application Architecture Modeling*, will help you gain the experience and confidence required for building models. This chapter will also provide you with more focused metamodels and more application architecture artifact samples that you can develop, to address the first scenario.

- *Chapter 6, Modeling in the Technology Layer*, will help you get familiar with the technology architecture layer and will show you how to model the physical layer of the enterprise, which is of interest to the CTO as indicated in the second scenario.

- *Chapter 7, Enterprise-Level Technology Architecture Models*, will continue on the journey toward the technology architecture layer, but we'll be focusing on its behavior and how to model it.

- *Chapter 8, Business Architecture Models*, will address parts of the third scenario by developing business architecture artifacts and focused metamodels.

- *Chapter 9, Modeling Strategy and Implementation*, will continue on the third scenario but with more attention on how to build strategic and implementation models and focused metamodels.

- *Chapter 10, Operating the EA Repository*, will show you some tips and guidelines that you can use to maintain your repository and keep it as the trustworthy source of information.

- *Chapter 11, Publishing Model Content*, will give you some tips and guidelines on publishing content so you can share your EA artifacts with non-Sparx users.

Now let us explore the first scenario and extract the user stories and tasks from it.

First scenario – application architecture

The first project that you will be involved in is the development of a new mobile application that can enable tracking shipments from the time they leave ABC Trading warehouses to the time they get delivered to retailers' warehouses. It will be a mobile application that truck drivers will install on company-provided smartphones that gives the location of the truck at any moment in time.

The reason for choosing an application architecture example is because many business solutions today involve the use of business applications. Whether your organization prefers to build or buy solutions, knowing how to define an application architecture and document it, will be required in all cases.

We will describe the scenario in more detail by putting more context around it, then we will extract the user stories and tasks to build the artifacts backlog. Let's start by looking more closely at the problem.

Scenario description

Today, ABC Trading retail partners use a web application on the ABC Trading website known as the *Trading Web* to check items' inventory and place orders online. Each retailer uses a user ID and a password for authentication before they can place orders. Payments can be made online as well as by checks in the mail. Retailers can check their orders' status to know if they have been processed, shipped, or are having issues, but they cannot know the location of their shipment at a point in time. For some retailers, shipments can take days and possibly weeks before arriving, depending on the distance, traffic, weather conditions, and many other variable factors, which makes it difficult for them to give accurate delivery dates to their customers.

Within the scope of this phase of the project, the targeted mobile application – which we will name the *Tracking App* – is supposed to provide the shipment tracking functionality only. However, it will be good to see other features being migrated from other ABC Trading applications to the mobile application in upcoming phases in the future but that requires different scoping.

The Trading Web application is aging in terms of technology, but it is still doing the essential parts of serving the business in terms of functionality and performance, so ABC Trading has no real requirement to replace it in the near future. The Trading Web application has some modern architecture patterns in it such as modularization and **Application Programming Interfaces** (**APIs**) for providing or receiving information. It is not comparable to modern shopping websites in terms of technologies and architecture, but it is not as obsolete as a monolith mainframe system.

Now we have sufficient information about the problem, we will extract the user stories from it in the next subsection.

Artifacts backlog

Your task as an enterprise architect is to write an **Application Architecture Document** (**AAD**) in which you will describe the targeted Tracking App in terms of high-level designs. This document will be used to guide the design and development of the application either internally or externally, so its content needs to be ready to be part of a **Request for Proposal** (**RFP**) or similar software procurement documents.

Our targeted audience of the AAD is the solutions architect who will use it as a high-level design and requirements document, to guide and influence the way they will design and develop the application. There is no solid line that defines where the enterprise architect role stops and where the solutions architect role starts as they overlap in many areas. But in general, the design details, the development of programming user stories, system requirements, test cases, and development schedule will be taken care of by the solutions architect and the development team so the AAD needs to stay at a conceptual-to-logical level of detail that is equivalent to **business use cases**. The AAD needs to contain at least the following sections:

- A definition of the business requirements that translates the problem description section into formal requirements

- A definition of what the target application shall do in terms of services and functionalities that users and other applications can use

- A definition of integration requirements with the existing systems

- A definition of the data that will be used in or produced by the application

- A definition of the required infrastructure technologies that will support the application

Let's see how many user stories and tasks we can extract from the previous requirements.

User story 1

To start with, we need to restate the problem definition in the form of a user story or stories. Our target audience (stakeholder) is a solutions architect, so the user story can be stated and described as follows:

As a solutions architect, I need to have an AAD describing the targeted Tracking App at a conceptual level that identifies the **Requirements**, **Application Services**, **Application Interfaces**, **Data Objects**, and **Technology Services**, so that I can build the Tracking App mobile application to realize the identified requirements and design principles and constraints.

Do not worry about the definitions of the keywords now as we will provide the TOGAF® and ArchiMate® definitions of each in addition to our elaboration and examples, all in the upcoming chapters. For now, let's build the artifacts backlog:

1. First, we need to provide a brief description of the Tracking App and who will be using it, which means that our first task is developing the **Application Component** context diagram.

2. After that, we need to build the requirements list and bear in mind any existing design principles and constraints that can influence the application, which means we need to define the **Application Requirements** catalog.

Then, we need to describe what this application component will provide to its context, and what it will receive from or provide to other applications. This can be translated into the following three tasks:

1. Define the application services catalog for the Tracking App.
2. Define the data objects catalog for incoming and outgoing data.
3. Define the application interfaces in which integrations with other applications will take place.

    Finally, we need to describe what technology (or infrastructure) services are required to run and operate the application. Remember that we do not have to name the actual products that will provide the technology services at this level.

4. We only need to identify what services are needed, so in other words, we need to define the technology services catalog for the Tracking App.

Since our EA repository in Sparx is a **green field**, which means there are no formal guidelines that we need to follow when developing our artifacts, we need to build these guidelines as we build and develop the catalogs. Please remember that every artifact needs to be based on a metamodel to tell the architects what elements and what relationships can be included, as we have described in the previous chapter. It will help in maintaining the integrity and consistency of your EA repository. With that said, we need to develop six focused metamodels, so each will have a task under the user story:

1. Build the application component focused metamodel.
2. Build the application requirements focused metamodel.
3. Build the application services focused metamodel.
4. Build the application interfaces focused metamodel.
5. Build the data objects focused metamodels.
6. Build the technology services focused metamodels.

Based on the size of the organization that you work for in real life and the complexity of its systems, fitting each of the preceding stories into a two-week period may not be possible, so you would break them down into smaller stories, but for the sake of the example, we will keep them as they are.

Remember that we are just beginning to build our EA repository so it is empty and we need to define the processes that are required for creating, modifying, and deleting artifacts, and this will be our second user story.

User story 2

The other thing that we need to define and build within our repository is the operating processes of the repository itself. We do not have to be sophisticated and detailed at the beginning because this may result in the scope creep that we described in the previous chapter. At the same time, we need to put in place some guidelines and documented processes of how to do things right.

Since the operating processes will be used across all the user stories that we have defined and will be defining in the future, it is a good idea to have a separate user story for each of the operating processes with a separate set of tasks.

As a lead enterprise architect, I need to define the proper guidelines and processes for creating, modifying, and deleting artifacts from the EA repository, so that every other contributing architect, including myself, can follow them to maintain the integrity and consistency of all EA artifacts.

The user story has already identified three of the tasks that need to be added to the backlog:

1. Define the process of EA artifact creation.
2. Define the process of EA artifact modification.
3. Define the process of EA artifact deletion.

Note that the term *artifact* is not limited to the diagrams only but includes the metamodels and the operating processes that are created as part of architectural work as well.

We have defined two user stories for the application development scenario, and it could be any application development project. Let's move ahead to the second scenario and create more user stories and tasks out of it.

Second scenario – technology architecture

With a tough year like 2020, ABC Trading is planning for a 20% budget cut, and many aspects of the organization, including the IT expenditures, must be re-evaluated and reduced to meet the new budget. IT expenditures are categorized into two main categories: **Capital Expenses (CapEx)** and **Operational Expenses (OpEx)**. CapEx are the expenses on the purchase and acquisition of new assets such as new hardware, software, or owned facilities. OpEx on the other hand are the periodical and continuous expenses that occur as a result of operating IT resources such as human resource salaries, subscription-based software and infrastructure, utility bills, rented facilities, and other similar types of *expenses*.

This example has been chosen to illustrate a simple example of changes that can occur at the **technology layer**, which in your case could be a cloud migration project, an upgrade to the existing technology stack, or the shutting down of an entire facility. The artifacts might differ slightly based on requirements but the idea is the same.

The next subsections will describe the scenario in more detail and will create more user stories and tasks and add them to the artifacts backlog.

Scenario description

The CTO has decided to analyze the existing IT assets, including all the software and hardware components, and identify whether there is any possibility to replace, merge, eliminate, or reduce the number of licenses of each in order to meet the cost reduction targets. Components that provide similar functionalities need to be identified and evaluated for possible elimination or merging. Components that can be replaced by other components performing the same functionality but at a lower price need also to be identified. Deprecating unused or rarely used components can also reduce the OpEx, and replacing some of the on-premises components with the equivalent cloud components is an additional option to consider.

The CTO would also like to have a dependency report showing which application uses which IT asset to help in analyzing the impact of making any changes to the existing technology stack.

> **Note**
> Please remember that this scenario has been created to demonstrate how to create EA artifacts in Sparx and is not meant to be used as a guideline to reduce IT budgets in any way. There could be better and more efficient ways to reduce IT budgets than the ones used in the example.

Now we have our problem defined and our requirements stated, our next step is to extract more user stories out of them and add them to the backlog.

Artifacts backlog

The CTO is looking for a document addressing the following list of requirements:

- Identify the existing IT assets including hardware and software components.
- Identify the services provided by each of the components and find any duplicated services.
- Identify the dependencies of applications on the existing IT assets to estimate the impact of changing, merging, or replacing any of them.
- Provide a **To-Be technology architecture** showing the proposed changes.
- Provide a plan for implementing the proposed changes.

From the preceding requirements, we can see that two user stories can be defined: an **As-Is technology architecture**, and a **To-Be technology architecture** with a list of **gaps** and a **roadmap**. The gaps will identify what is missing between the As-Is and the To-Be architectures, while the roadmap will identify the set of initiatives and projects that will be needed to close these gaps.

> **Note**
>
> Gaps do not always indicate something necessarily is missing from the As-Is that needs to be added to the To-Be. Gaps indicate differences between two states of the architecture regardless of which state has more elements than the other. If the As-Is has more components than the To-Be, that is still considered a gap even though we need to remove elements to bridge it.

We already have two user stories in our backlog so the next will be user story 3.

User story 3

The third user story is an **As-Is technology architecture**, described next.

As CTO, I need to have a list of all the technology assets in the enterprise, know what they provide, and identify what application uses which asset so that I can identify the redundant services and know the impact of replacing or retiring some of these assets.

As mentioned in the previous scenario, we can leave the formal definitions for *Section 2, Building the Enterprise Architecture Repository*, when we start creating the required artifacts and pay more attention to building the artifacts backlog for now. To start with, we need to translate the CTO terminology into the equivalent EA ones. Lists are **catalogs**, and the technology assets are the hardware and software that support the enterprise, so architecturally they are called **technology nodes**. These technology nodes provide **technology services** to the enterprise including the **application components** that depend on them.

> **Note**
>
> It is very important for an enterprise architect to listen to stakeholders' requirements in their own language and to not attempt to enforce EA acronyms on them. It is your job to make that translation, not theirs, because you are the one bridging the communication gaps across the enterprise, not creating new gaps.

Having the translation part done, the third user story will have the following tasks:

1. Define the technology nodes catalog.

2. Define the technology services catalog.

3. Additionally, we need to show the relationship between these two catalogs, so we will define the technology-nodes-to-technology-services matrix.

4. We also need to define the relationships between application components and technology nodes, so we need to define the application components catalog.

5. Then we need to define the application-components-to-technology-services matrix to build the relationship between the two catalogs.

As mentioned earlier, since our EA repository in Sparx is brand new, we need to build the metamodels that guide the development of each artifact. Therefore, we need three focused metamodels, one for each:

1. Build the technology node focused metamodel.

2. Build the technology service focused metamodel.

3. Build the application component focused metamodel.

Notice that *user story 1* has a task for building the technology service focused metamodel so most probably this is a duplicate of the same task. Let's keep both in the backlog for now and decide which one to remove later.

User story 3 looks good now with sufficient details and tasks to start working on it, so let's move on to the next user story.

User story 4

The fourth user story is about defining some To-Be artifacts describing the targeted technology architecture with the proposed changes and a plan for implementing them.

As CTO, I need to have a list of the duplicated services and those services no longer being used, along with the possible impact of deprecating, merging, or replacing the technology assets that provide them, so that I can create an action plan for implementing the changes.

Since we have already developed the technology services and technology nodes catalogs in the previous user story, we need to utilize them in this user story to derive smaller but more specific catalogs out of them. The duplicated and unused technology services catalogs are subsets of the main technology services catalog:

1. Define the duplicated technology services catalog.

2. Define the unused technology services catalog.

    Then we need to define what it looks like if we remove, merge, or replace these services.

3. Define the To-Be technology services catalog.

4. Define the To-Be technology nodes catalog.

    Define the **To-Be application components** to **technology services** matrix so we know the impact of making these changes on the existing applications' gaps between them and build a plan for bridging them. Therefore, we need to add two more tasks to the user story.

5. Define the **technology gaps** catalog.

6. Define the **work packages** catalog.

    And of course, define the mapping between the gaps and the work packages so we need the following.

7. Define the **work packages** to **technology gaps** matrix.

    Think of work packages as a group of actions that need to be addressed to resolve a common problem. They can be whole projects or smaller components of projects. Since we already have tasks for building the technology nodes- and technology services focused metamodels, we do not have to repeat them here. We will need, however, to build the focused metamodels for the gaps and the work packages:

8. Build the gap focused metamodel.

9. Build the work package focused metamodel.

Do not forget that we need to define the governance model for the artifacts created in *user stories 3 and 4*. Looking back at *user story 2*, we can see that its tasks are generic for any artifact so there is no need to duplicate or customize any of them.

Now we have four user stories in the backlog, we can move on to the last scenario and extract more stories out of it.

Third scenario – business architecture

ABC Trading has always been a wholesale company and has never sold any items directly to end consumers. The relationship with their retail partners has been excellent and nothing has emerged to alter the – so far successful – business model. However, the only constant thing in this world is change and major changes occurred in the world in 2020 that affected businesses no one thought would ever be affected by crises – of course, I am referring to the COVID-19 pandemic and the lockdowns of cities and countries.

This scenario is about a business problem that needs to be resolved not only with technology but with **people**, **business processes**, **information**, and **technology**, which together enable your business to provide services to its consumers. In other words, these four aspects combined make your organization *capable* of providing services. Therefore, it is called **business capability modeling** and it is an immensely powerful business planning tool that helps you to understand how changes to the provided services require changes to the components that support them and vice versa.

You can map this example to any similar changes to the business of your organization – if your business is in manufacturing and wants to introduce a new product or possibly to discontinue an existing one, or if your business provides financial management services and is planning to introduce a consulting line of business, you can use the exact same concept that will be provided here.

The following subsections will provide more details of the scenario and will extract new user stories and tasks and add them to the artifacts backlog.

Scenario description

The COVID-19 pandemic has changed the way people work, study, and shop. People used to enjoy going to shops to buy what they need, but with everyone staying home, the demand for online shopping and delivery services has increased dramatically and has made clear the difference between a successful business and a struggling business.

ABC Trading's retail partners were no exception from the effects of the pandemic and some of them went out of business because they were not ready to change. ABC Trading's revenue went down by 20% as a result, and all the credit for it remaining in business goes to the retailers with established online shopping channels. With that said, ABC Trading has decided to establish online retail sales to sell goods directly to customers and reduce its dependency on retail partners. A feasibility study was done a few months ago and showed promising profit figures, so the CEO has decided to proceed with the initiative. Part of the proposed risk mitigation plan is to reuse or reallocate the existing resources if possible before considering procuring new ones.

The CEO has asked you to develop a high-level business architecture document describing the new online sales service, what ABC Trading needs to do to start providing the service, how much can be reused from the existing resources, what is missing, and what needs to be built and developed.

Artifacts backlog

To start with, we need to break down the CEO's requirements:

- Define how the new business service will be provided.
- Identify how the existing business services are provided and find the components that can be reused for the new service.
- Define the gaps that need to be closed to be able to provide the new service.
- Define a roadmap with the required actions to take.

The next step is to build user stories based on these requirements. Two user stories can be identified.

User story 5

This user story is all about business capability modeling for the new service.

As CEO, I need a model that shows what we need to be able to provide the online shopping service to customers so that I can identify the required changes to make.

As we mentioned in the previous two scenarios, do not worry about the definitions now – we will introduce them when we define the artifacts. For now, we must continue building the artifacts backlog, so we need to identify the required tasks under this user story but using architectural terminologies:

1. Build and develop the To-Be **business capability model** for the online shopping **business service**.
2. Build the To-Be **business actors** catalog that defines what human resources are needed for providing the service.
3. Build the To-Be **business processes** catalog that defines how the new services will be provided.
4. Build the To-Be **application services** catalog that defines the required application services.

5. Build the To-Be **data entities** catalog that defines the required data and information.

6. Build the To-Be **technology services** catalog that defines the required technology services.

We already have some user stories for some focused metamodels, so we need to add the following ones:

1. Build the business capability focused metamodel.

2. Build the business service focused metamodel.

3. Build the business actor focused metamodel.

4. Build the business process focused metamodel.

5. Build the data entity focused metamodel.

Now we can add *user story 5* to our backlog and move on to *user story 6* to find what components we can reuse to provide the new service.

User story 6

User story 6 is a continuation of *user story 5* but because the latter was already too long, it needed to be split into two stories instead of one.

As CEO, I need to know what the closest existing services to the new ones are and what their components are, so that I can identify what can be reused and what actions need to be taken.

The architectural tasks for this user story are as follows:

1. Identify the **As-Is business capability model** for the selected business service.

2. Identify the **As-Is business actors** catalog that defines what human resources currently provide the current service.

3. Identify the **As-Is business processes** catalog that defines how the current services are provided.

4. Identify the **As-Is application services** catalog that defines what application services support the current business service.

5. Identify the **As-Is data entities** catalog that defines the data and information that are required currently for providing the service.

6. Identify the **As-Is technology services** catalog that defines what technology services support the current business service.

7. Identify the **gaps** catalog between the **As-Is** and **To-Be** models.

8. Identify the **course of actions** catalog to define required actions to take to be able to provide the new service.

   As with the other stories, we need to build a focused metamodel for the artifacts that we are planning to produce. The metamodels can be used for both the As-Is and the To-Be artifacts so we do not need to repeat the tasks that were defined earlier.

9. The only new element that does not have a focused metamodel is the **course of action**, so we need to build the **course of actions** focused metamodel.

All the required governance processes have been identified in *user story 2* and there is no need to change or customize any of them to fit the additional scenarios now, so we will keep all of them as they are defined in *user story 2*.

With a backlog like the one we have defined, it can take any time between 6 months and a year depending on the size of the organization. Let's recap and summarize what we have learned from this chapter before moving on to building the EA repository in *Section 2, Building the Enterprise Architecture Repository*.

Summary

In this chapter, we have introduced the three scenarios that we will use to build the EA repository in Sparx. Every scenario and story mentioned in this chapter is purely fictional and was written to illustrate the different roles an enterprise architect can play in different projects and engagements at all the layers of the enterprise at the same time. We have learned how to build an artifacts backlog that contains the user stories to address stakeholders' requirements. We have also learned that defining metamodels is as important as building the artifacts, because without them we would not be able to maintain consistency across the EA repository.

In the next chapter, we will start building the EA repository using Sparx and populate it with the content that addresses all these user stories. The fun part is just about to begin.

Section 2:
Building the Enterprise
Architecture Repository

This section provides practical examples of the different types of artifacts that an enterprise architect can develop to build an architecture repository while being involved in different types of activities with different business units.

Building an enterprise architecture repository involves building a library of reference models that architects can use to build a library of artifacts (mainly diagrams). Each artifact can tell a piece of information about your enterprise, and by combining multiple interconnected artifacts, you are able to see the bigger picture. Being able to easily navigate between artifacts in different architecture layers will help you understand and trace how each element is connected and to what. Additionally, having a library of nested artifacts can show or hide information as needed at different levels of detail, by drilling up or down between diagrams.

This section comprises the following chapters:

- *Chapter 3, Kick-Starting Your Enterprise Architecture Repository*
- *Chapter 4, Maintaining Quality and Consistency in the Repository*
- *Chapter 5, Advanced Application Architecture Modeling*
- *Chapter 6, Modeling in the Technology Layer*
- *Chapter 7, Enterprise-Level Technology Architecture Models*
- *Chapter 8, Business Architecture Models*
- *Chapter 9, Modeling Strategy and Implementation*

3
Kick-Starting Your Enterprise Architecture Repository

When stakeholders introduce an idea of developing a new application to solve some business problems, your job as an enterprise architect is to help them and everyone involved to have a better understanding of what the application is for. Long before starting any designs, you need to identify what service the application is intending to provide, who will be using it, and with which existing applications it will exchange data. Whether you are planning to build this application in-house, hire a contractor to build it, or buy it off the shelf, you still need to develop these conceptual models to help the stakeholders make the right decision – including the decision of approving or rejecting the application altogether. Your role as an enterprise architect is to clarify and formalize the ideas and turn them into requirements for the solutions architect to build and/or deploy the application.

We have a backlog from the previous chapter that contains tasks for creating models (diagrams) and tasks for creating metamodels. The words **diagram** and **model** will be used interchangeably as they both have the same meaning. Architects prefer to use model while users prefer to use diagram, so keep that in mind. **Metamodels**, in a nutshell, are models that tell us how to build models, so it makes sense to start by building the metamodels first and then derive models from them. However, in some cases, you may find that starting from the model and reverse engineering it to develop the metamodel is more convenient, especially if you are experienced in reading and interpreting the standard metamodels of TOGAF® and ArchiMate®. It is the ever-lasting question of what comes first, *the chicken or the egg*.

In this chapter, we will start with the model and will do the focused metamodel in the next chapter. This means that we are starting with the *egg* on purpose as we do not want to spend a lot of time on the theory (the metamodel part) without you experiencing an example of what the result is. We will learn the following while practicing building the Tracking App application context diagram:

- Building the application component context diagram
- Establishing your first diagram
- Adding elements to the diagram

If you are new to using Sparx as an enterprise architecture tool, this is your chance to learn the basics by using an example and following step-by-step instructions, so let's get started.

Technical requirements

If you have not installed Sparx Systems Enterprise Architect yet, it is now the time to do so; we will use it here and in all the remaining chapters of this book, and it is better to practice using it while reading. If you do not have a licensed copy, you can download a fully functional 30-day trial from the Sparx Systems website (`https://sparxsystems.com/products/ea/trial/request.html`). If you prefer to use another tool, you can do so, but you will need to figure out how to implement the examples in your tool of choice.

Installing Sparx is fairly straightforward for desktop users, as explained in this online video: `https://sparxsystems.com/resources/show-video.html?video=gettingstarted-installingea`. Corporate users need to consult their network admins for possible restrictions on installing software, and they will also need to get the proper connection string to the backend SQL database.

Because this is the first chapter to use Sparx Systems Enterprise Architect, we will be delving into a large amount of detail around navigating the tool. If you're already familiar with Sparx, please be patient. Subsequent chapters will go into less detail as we expect everyone to have become familiar with the tool by then.

Building the application component context diagram

A **context diagram**, as the name implies, is a diagram that shows the *surroundings* of a specific component. The main purpose of context diagrams is to provide a high-level introduction to an element such as an application component by answering the following questions:

- What is the element and why does it (or will it) exist?
- What does it provide to the enterprise?
- Who will be using it?
- Are there any other applications that interact or exchange data with it?
- What data is being (or will be) exchanged?

Context diagrams can be useful for describing existing elements (as is) and for describing targeted elements (to be). The same concept applies, and the same types of elements can be used in both cases. Before you roll your sleeves up, let's see how to establish your first diagram.

Establishing your first diagram

In this section, we will be building a brand-new enterprise architecture repository that will contain all our diagrams and elements. Diagrams visually show how elements are related to convey an idea, such as what a specific component is composed of, how a specific service is provided, or how a group of data centers are connected and what they contain. An element can be represented by different diagrams, each showing a different aspect of it, so the same element can appear in many diagrams.

Packages, on the other hand, represent a physical containment of their content, such as elements, diagrams, and other packages. A child element can have one parent only, but parents can have many children. Packages look and act like folders in file systems as they *contain* other elements, while elements and diagrams are the equivalents of files in file systems. You can nest packages within packages as much as you may need, but you need to keep in mind that very deeply nested packages can make it difficult for you and other repository users to find the information, so keep the users in mind when you build the structure.

If users find it easy to locate and get information in the enterprise architecture repository, they will use it as a reliable source of information; otherwise, they will keep doing whatever they are comfortable with. Therefore, make sure that the structure of the repository, including the names of the packages, elements, and diagrams, makes sense. As we progress in this book and our repository grows and matures, the need for restructuring will start to become clear, and we will see how we can respond to these needs. There is no one right way to do things, of course, but I am providing you with a tested way.

The repository itself is a relational database. If you store it locally on your machine, the database will be of the **Microsoft Jet** type (**Microsoft Access** basically), but it will be accessible only by you. You can choose to create the repository in a centralized server using any number of different database engines. All Sparx instances on desktop machines will connect to it and contribute to the enterprise architecture content.

> **Note**
>
> If you are using Sparx within an organization, you may need to consult your IT helpdesk to get the connection string to the SQL server, and there might be some rules, constraints, and guidelines that you need to be aware of.

Migrating content between your local database and the centralized database can take place at any package level, so you can export only the new content of the package that you need. From the Sparx application, navigate to the **Publish** ribbon. From the **Model Exchange** section of the ribbon, select **Export-XML**, then **Export XML for Current Package**; read or bookmark the related online help for more information on how to do this (`https://sparxsystems.com/enterprise_architect_user_guide/15.2/model_publishing/exporttoxmi.html`). We have just started a brand-new repository, so we do not need to worry about migrating content now, but one day, you will need to move content between different repositories for sure, so you know now that it is a simple thing to do.

In this section, we will learn the following:

- How to create the repository file
- How to create the diagram
- How to describe the diagram
- How to change the diagram theme

This section is our first step toward building our enterprise architecture repository, so make sure that Sparx is up and running. Now, let's get started.

Creating the repository file

After starting Sparx, you will be welcomed by the start page, which asks you whether you want to open a file, create a new one, or open one from the recently opened projects. This is our first diagram in a brand-new repository, so we need to create a new project:

1. Choose the **Create a New Project** option under the **New** menu and select a folder to place the Sparx project file in.

2. Type any name of your choice; I will use the name `EA Repository` for this purpose.

   You should now see a single item named **Model** in the Project Browser window. This is called the **root node** and we will use it as our top-level package for our architecture content for now.

3. We need to give the root node a more meaningful name than **Model**, so click on the root node then press *F2* to rename it to `Architecture Content.`

The structure of the repository is extremely flexible when there is a need to change it, and this need will come sooner or later. The enterprise will keep changing and the way you understand it will keep changing too, so the repository must be flexible enough to accommodate these changes with ease. Fortunately, this can be easily done in Sparx, and it is similar to organizing files and folders in a file system, so do not worry if you started with a different understanding in mind and want to change it later.

For example, if you put an element in package A, link this element to other elements, place the element on several diagrams, and then decide to move it to a different package, the links will not be broken no matter where the element is relocated. Using an enterprise architecture tool such as Sparx versus using simple Microsoft Office documents can have a huge positive impact on your ability to successfully implement an enterprise architecture practice. If your enterprise architecture artifacts are all in Office documents stored in a file system or even in a document management system, any links between these documents (such as reference URLs) will be broken the moment you start moving files around, and it can be very costly to fix. Remember that everything is connected in the enterprise, so having broken connections can be fatal to the reputation of the enterprise architecture repository. In *Chapter 11*, *Publishing Model Content*, we show you how to render your enterprise architecture repository information in different formats, including Microsoft Office documents.

We have an empty repository, so now we need to create a new package to contain our first diagram and all the related elements.

Creating the diagram

We need to create a package to contain the diagram that we will create and that will contain all the elements that will be placed on it. Diagrams cannot be created under the root node directly; they need to be created within a package under the root node or under another package.

Creating packages and creating diagrams can be performed together in one step or separately as two steps. In this example, we will create both in one step, and I will show you in other examples how to create each in its own step.

To create a package, follow these steps in order:

1. Click on the **Architecture Content** root node and from the Toolbar, choose **Design > Model > Add > Package** to create a new package. This will bring up the **New Package** pop-up window, as in the following figure:

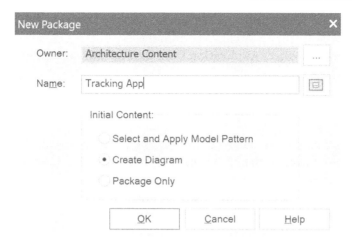

Figure 3.1 – New Package pop-up window

2. Enter Tracking App in the **Name** field and select **Create Diagram**. Click **OK**, and the **New Diagram** window will open, as shown in *Figure 3.2*.

    If you selected **Package Only** in the **New Package** window, the **New Diagram** window will not be launched automatically, and you will get an empty package. You can add a diagram at any time by right-clicking on the package and selecting **Add Diagram** from the context menu, which will bring up the same **New Diagram** window, as follows:

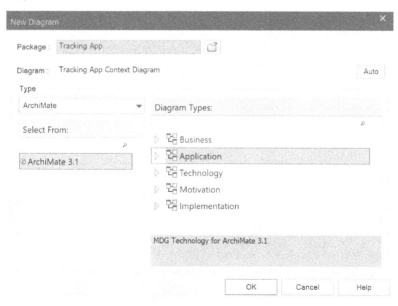

Figure 3.2 – The New Diagram window

3. Type `Tracking App Context Diagram` into the **Diagram** field, select **Enterprise Architecture** > **ArchiMate** from the **Type** drop-down list, highlight **ArchiMate 3.1**, and select **Application** from the available diagram types.

4. Click **OK** to accept the selections, which will close the window and open a blank diagram for you to start drawing on.

We need to pause for a few seconds here and explore what we have on the screen after the previous actions. As you can see, an empty diagram is now shown in the diagram area as a **tab**. Clicking on the **x** icon on the tab will close it, and you can re-open it from the project browser window. In the project browser area to the left of the screen, you can see the newly created diagram under the **Tracking App** package. If you *double-click* the diagram icon, the diagram will be reopened and displayed in the diagram area again, as shown here:

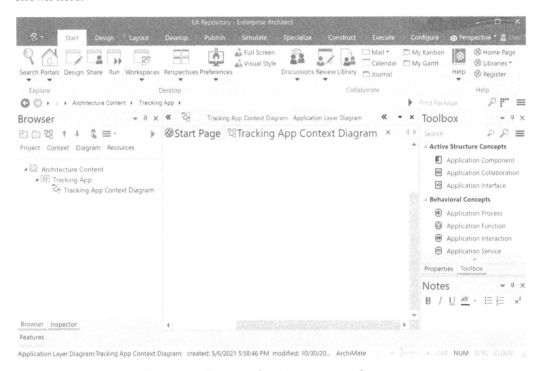

Figure 3.3 – Empty application component diagram

To the right, there are two areas: the **Toolbox** and **Notes** areas. The **Toolbox** area changes according to the type of diagram that you have chosen in the **New Diagram** window. Because we chose **Application** as the diagram type, the **Toolbox** contains all ArchiMate® 3.1 Application-layer elements.

The **Notes** area is context sensitive to the currently selected item. If you select a different element from the Project Browser window, the **Notes** content will change accordingly. We will use the **Notes** area to describe all the elements that we create and make the content ready for publishing.

Describing the diagram

It is a good practice to add descriptive text for every diagram and every element you create. What you think of as obvious information may not be for other users, and sometimes may not be obvious to you if you revisit a diagram that you have created after a while. Adding comments in code while programming is just as important for helping yourself when you come back to it as helping other developers who will be working on it.

In this section, we will see two different ways to add descriptive text to your diagrams: one is by using labels and the other one is by using the diagram notes fields.

Using labels

Labels act like titles to diagrams and they immediately inform the readers what the diagram is about. Labels are elements, so we need to add them from the **Toolbox** to the diagram area and add descriptive text to them. Follow these steps to add a label to a diagram:

1. Scroll down the **Toolbox** all the way near the bottom until you find the **Common** section.

2. Expand the **Common** section and find **Text Element**. Click and hold **Text Element** and drag and drop it onto the diagram. Alternatively, you can click once on **Text Element** (or any element in general), move the mouse where you want to create it, and click on the diagram to create the element where you clicked.

3. Type `Tracking App Context Diagram` in the designated text area and click anywhere on the diagram to see your text element updated.

4. You can format the font and adjust the font size by locating the font configuration button in the **Layout** > **Style** menu, which is indicated by a capital letter **A**.

5. Select the font name, style, and size that you like from the pop-up window. The defaults are **Calibri**, **Regular**, and size **8**. I will keep using the same font name and style but will change the size to be *16 pt* in labels and *10 pt* in all the other elements.

6. Since this is a label, change the size to **16** and press **OK** to accept and close.

You can move the label and place it at any desired location on the diagram, but labels are usually placed at the top-left corner of the diagram area as this is where readers will usually start reading.

> **Note**
> Some organizations are required by law to use specific font names, sizes, colors, or styles in their content, especially public-facing ones. Some other organizations enforce using specific font styles for marketing and brand identity purposes. You need to check with the designated units in your organization for any such requirements and make sure that you follow them in the entire repository.

Keep in mind that text elements can be used for any type of text content that we want to add to the diagram. They can contain a description of something, a set of instructions, side notes, things to do, and so on; they have many different usages. Using them as labels is just one of them.

Using the diagram notes field

Diagram notes are fields that can contain any text that you feel is useful to add. They are not visible in the diagram area, but Sparx users can see them in the **Notes** window when they click on the diagram area. Another useful benefit of entering a diagram description in the **Notes** field is making the documents that will be published from Sparx richer in content. You can customize your document templates in a way that prints the **Notes** field before or after the diagrams, as we will see later in *Chapter 11*, *Publishing Model Content*.

To add a description in the diagram **Notes** field, we need to do the following:

1. *Right-click* on the diagram background and select **Properties**, which will open the **Properties** pop-up window for the diagram. As you can see, there are multiple tabs in this popup, and we will explore some of these options as we go. On the first tab, labeled **General**, you can rename the diagram, change the author's name, change the notes, and adjust other settings.

2. In the **Notes** field at the bottom, type any text that describes what this diagram is about. You can use this sample text if you want: *This diagram shows the context of the Tracking App application component. It describes what the application component is, what it provided, who will use it, what other applications will integrate with it, and what it needs in order to operate.*

3. You can educate your Enterprise Architect users to look at the **Notes** section of diagrams if they want to know what the diagrams are about and what viewpoint they are addressing. Make your notes short and descriptive because most of the description should be conveyed in the diagram that you are creating, not in a text note.

Keep this popup open as we need to change the theme of the diagram in the next step.

Changing the diagram theme

Themes are predefined settings that affect the way your diagrams look, mainly by applying different background and foreground colors. Themes are applicable at the diagram level, so you can have different themes for different diagrams.

However, because the selection of themes affects the way your diagrams look, I advise you to stick to one theme for the entire repository. Every enterprise architect who contributes to the content will be required to use the same theme to maintain consistency. If you think that this will be too much to enforce, then you're better off using the default theme and asking everyone not to change it.

To change the theme of the current diagram, you need to have the diagram properties popup open:

1. Go to the **Theme** tab and scroll through the dropdown until you find **High Contrast White**.

    This is my favorite theme because it is the closest one to how colors appear on paper. It reduces the possibility of seeing things differently on paper than on screen.

2. Select **High Contrast White** from the list and click **OK** to apply the changes, and then close the popup.

The choice of themes and colors is purely up to your preference, so you are free to pick any one of them you feel comfortable with.

> **Note**
>
> If you are working with a team of architects, you all need to agree on the same theme and use it to maintain consistency among your deliverables. If team members are using different themes but not using the default colors of that theme, the way diagrams will look will vary from one computer to another, which affects the consistency among enterprise architecture team members.

We have established the initial structure for the content that will be created and established an empty diagram, and in the next section, we will add more elements to it.

Adding elements to the diagram

Element is a generic word for anything you place on a diagram. New elements are provided in toolboxes, and for each type of diagram, Sparx makes one toolbox the default toolbox for that diagram. When creating our diagram, we chose that we wanted to create an ArchiMate® 3.1 Application diagram; therefore, Sparx has set ArchiMate® 3.1 Application as the default toolbox for us. This can be changed, as we will see later, but first, we need to add the main element to the diagram, which is the application component.

Starting with the application component

It is always a great idea to have an imaginary picture of the diagram in your head or a sketch on a piece of paper before starting to model it. In this diagram, we need to convey what the new application is about, what it provides, who will be using it, and what other applications need to integrate with it. Therefore, I am imagining an application component in the middle of the diagram, surrounded by Actors and applications, each getting or sending something to the application. This should be a conceptual diagram that describes the application at a high level, and it should be easy to read and be understood by any audience.

Defining an application component

An application component is defined as "*an encapsulation of application functionality aligned to implementation structure, which is modular and replaceable*" (https:// pubs.opengroup.org/architecture/archimate3-doc/chap09.html#_ Toc10045392).

An application component is meant to provide one or more *functions* to users and other applications (formally known as **Actors**). It *encapsulates* these functionalities, so it is unknown to the Actors how it works internally, but services are provided to them through interfaces. Interfaces can vary based on the Actors, such as the **User Interfaces** (**UIs**) for communicating with human Actors (or *users*) and the **Application Programming Interfaces** (**APIs**) for communicating programmatically with other applications. In modern programming, UIs are nothing more than loosely coupled application components that communicate data captured in forms to other backend application components through APIs.

Application components can comprise other application components, so the size of the component does not matter if it is modular and replaceable. An application component can be as large as a monolith **Enterprise Resource Planning** (**ERP**) application that has hundreds of modules, each having hundreds of submodules, or it can be as small as a microservice that takes in a text value and returns it with double quotes around it. These are both considered application components according to the definitions.

ArchiMate® 3.1 provides two notations for modeling application components, rectangular and borderless, as shown in the following figure:

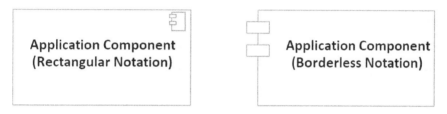

Figure 3.4 – Application component notations in ArchiMate®

That is all for the definition, and by now I hope that you have Sparx up and running because we will start modeling right away.

Creating an Application Component

The first element that we need to add to the diagram is the Application Component element, so follow these steps:

1. From the **Toolbox**, drag an **Application Component** element and drop it in the center of the diagram.

2. Sparx will expect you to rename the element to a more meaningful name, so you can type `Tracking App` and then press *Enter* or click anywhere outside the element.

You have performed two steps in one: you have created a new element in the repository that represents Tracking App, and you have placed that element on the diagram. There are other ways to create elements directly in the Project Browser without having them placed on a diagram, but this is the most common and easiest way to do it.

Keep in mind that if you want to create another diagram that contains Tracking App, you must reuse the same application element from the Project Browser. If you drag another Application Component element from the Toolbox and name it `Tracking App` too, Sparx will accept it and will never check for or warn you about possible duplicates because the name is not a unique identifier for elements.

You need to know whether you want to create a new element or reuse an already existing one. Always remember to drag and drop from the *Toolbox* to *create new* elements and drag and drop from the *Project Browser* to *reuse* existing elements.

If you are not sure whether an element exists or not, you can press *Ctrl + F* to search for it and reuse it. Searching for elements is essential to building a well-organized and trusted enterprise architecture repository, and luckily, Sparx has a good search engine that you will find handy. We will talk more about searching as we create more elements and diagrams in the repository that require us to search.

Adding a description to a component

Make it a habit to describe every element you create in the repository, because the enterprise architecture repository is supposed to be the source of the most accurate information about your enterprise elements:

1. Open the **Properties** window of the **Tracking App** element by right-clicking on the element on the diagram and selecting **Properties** > **Properties**. Alternatively, you can select the element and then press *Alt + Enter* as a keyboard shortcut to open the same **Properties** window.

2. Type in a short description of the application component in the **Notes** section, such as *This mobile app will be installed on truck drivers' smartphones to allow the shipping managers and partners to track the location of the truck at any given moment.*

It is better to have the information and not need it than to need the information and not have it. On the contrary, awfully long notes can discourage readers from looking at them, so you need to maintain a balance, which I suggest is between three and six lines of descriptive notes. There are other ways to add lengthy documentation to an element, which we will explore later.

Displaying a note in a Note element

After you close the **Properties** window, the notes will be hidden, and if someone is looking at a published or printed version of the diagram, they will not be able to see the notes. Therefore, we need to place a **Note** element to show this information:

1. Scroll down the **Toolbox**, expand the **Common** area, and you will see the **Note** element at the top of the list.

2. Drag the **Note** element and drop it onto the diagram area close to the Application Component element.

> Important
>
> Do not get confused between the **Notes** *field* and the **Note** *element*. Every element and every diagram has a designated **Notes** field that you can use to write descriptive text about that element or diagram. These notes are not visible on the diagrams by default but can be visible in some Sparx windows. On the other hand, there is a **Note** element, which is part of the **Common** group in every toolbox. You can place it on a diagram to display a descriptive message about anything you want, and it is always visible to the reader, just like a *Post-it* note.

Notice that adding the **Note** element to the diagram did not appear in the project browser; that is because some elements, such as **Note**, **Text**, and **Boundary**, are not meant to be reused and they belong only to the diagram that they have been placed on. In other words, they are not considered **architectural elements**, so they have no physical presence in the enterprise architecture repository, but just a visual presence on a specific diagram.

The **Note** element is empty, and you can type anything in it or link it to another element and display that element's **Notes** field instead. In our case, we need to display the text that we have typed in the application component's **Notes** field to make it visible always because our goal is to tell the readers what the application is.

3. Click on the **Note** element and you will see three small action buttons on the right side of the element: an arrow, a menu, and a brush, as you can see in the following diagram:

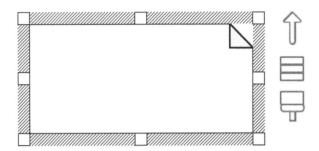

Figure 3.5 – Action buttons appear when you highlight an element

These three buttons provide contextual actions, and they are available for each element on the diagram. Click on the Application Component element now and you will see the same action buttons next to it:

- The *arrow*-shaped button is used to create *relationships* between elements.

- The *menu*-shaped button displays a *context menu*.

- The *brush*-shaped button provides appearance and *formatting* options.

You can still access all these options using different menus, so it is just a matter of convenience.

4. Click and hold the arrow button and drag it to the Application Component element. Once the mouse pointer is above the application component, release the mouse and you will get a list of possible relationships that you can create between the source (in this case, the **Note** element) and the target (in this case, the Application Component element).

5. For these two types of elements, there is only one type of possible relationship: **Link**. When selected, you will see a dotted line now connecting the two elements, which represents a **link relationship**.

The **Note** element is still empty, and we can type the same text description that we typed earlier in the application component's **Notes** field. However, this means that we must maintain that text in two different places now. This can be a time-consuming task when you have a large repository full of elements and notes, and most of the time, you will end up making changes in one place and forgetting the other, resulting in inconsistency.

What we need to do is to tell Sparx that we want to display the exact same application component notes in the **Note** element.

6. To do that, *right-click* on the link relationship and click on the **Link this note to an element feature** option, which will bring up a pop-up dialog box.

7. Open the **Feature Type** drop-down list and select **Element Note**.

8. Click **OK** to accept the changes and close the popup.

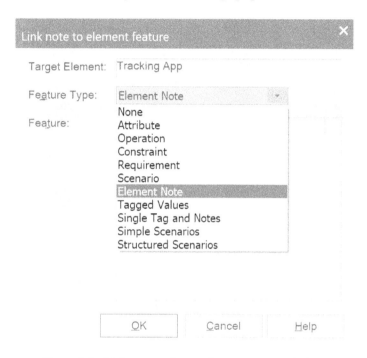

Figure 3.6 – Link note to element feature pop-up window

You can see now that the text that you typed in the **Note** section of the application component is now displayed in the **Note** element on the diagram.

Next, you will see how to style the elements that we have placed on the diagram so far.

Styling elements

You can optionally change the color of the Application Component and **Note** elements to give them a nicer look. Take the following steps to do that:

1. From the **Layout** > **Style** menu, set the fill color to *white* (RGB 255, 255, 255) and the border color to *royal blue* (RGB 65, 105, 225).

2. From the same **Layout** > **Style** menu, change the font size from **8** to **10**.

Since Application Component will be the main element on the diagram, let's give it a thicker border and make it bigger in size so the reader will be able to tell from the first look which element in the diagram is the main subject.

3. Using the same **Layout** > **Style** menu items, adjust the width of the border from the default value of **1** pt to **2** pt.

4. Also, change the fill color of the **Note** element to *light yellow* (RGB 255, 255, 224) to give it the style of a real *Post-it* note.

5. Finally, change the notation of the application component to the borderless notation by right-clicking on it and unchecking **Advanced** > **Use Rectangle Notation**. Alternatively, you can click on the Application Component element, then click on the brush action button and uncheck **Use Rectangle Notation** from the menu.

6. Save your changes by pressing *Ctrl + S* on your keyboard or click on **Layout** > **Diagram** > **Save** from the Toolbar.

By now, your diagram should look like the following diagram, or you should at least know why it does not. You can see the label in the top-left corner, the application component, and the **Note** element displaying what is in the **Notes** field of the component:

Tracking App Context Diagram

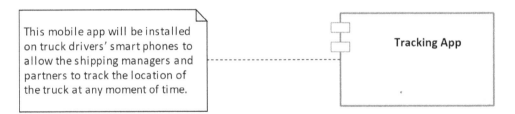

Figure 3.7 – The Tracking App context diagram so far

The next thing that we need to show on the diagram is what Tracking App provides to its users.

Introducing application services

Application components are meant to provide services to users, so we need to add more information to our diagram telling them what services Tracking App will provide. Based on the description of the first scenario in *Chapter 2, Introducing the Practice Scenarios*, Tracking App is targeted to provide a single service within the current scope, which is *Vehicle Tracking*.

The best way to identify the services that an application component provides is to refer to the **Use Case** documents if available. Use cases describe how Actors will interact with a system; they tell you what services a component exposes to the external Actors, which can either be users or other applications. Remember that we are still conceptualizing the application so we may discover more information as we share the diagrams with the stakeholders. A simple diagram will help stakeholders to reorganize their thoughts and visualize what they are expecting to see in the target solution, so high-level use cases are what you need to consider at this stage.

This subsection will show you how to model the application services that are provided by an application component.

Adding an Application Service element

We need to describe the services that are provided by Tracking App. ArchiMate® has provided the Application Service element for this purpose, and each service will be represented by an Application Service element.

Like what we have previously done when we added an application component and a note to the diagram, the Application Service element can be found in the same toolbox. Follow these steps to add the Vehicle Tracking service to the diagram:

1. From the **Toolbox**, drag an **Application Service** element, drop it on the diagram, and rename it to `Vehicle Tracking`.

2. Open the **Element Properties** window and in the **Notes** section, type a short description about this service, such as *This service provides information about the current location of the smartphone that will be carried by the truck drivers during shipment delivery.*

3. Click **OK** to close the **Properties** window and return to the diagram.

To remain consistent with other elements in the diagram, we need to style the Application Service element, which will be described next.

Styling the element

Optionally, style the Application Service element to give it the look and feel of other application architecture elements:

1. Follow the same steps that we defined for the application component to style the application service. Make the font size **10**, the fill color *white* (RGB 255, 255, 255), and the border color *royal blue* (RGB 65, 105, 225).

2. Since the application service is not the center of focus in this diagram, we need to keep the border width as **1**.

To avoid repeating the same styling steps for every element we add, it will be better if we save the style and reuse it whenever needed, which will be explained next.

Saving the style for reusability

We will be using the same element style over and over for every application architecture element in the repository, so it is best to save it for later reuse. Saving styles also increases the consistency between different team members as they will all use the same style. Follow these steps to save the style for reuse:

1. Highlight the Application Service element on the diagram and locate the **Layout** > **Style** > **Manage appearance styles** toolbar button (which has a disk and pencil icon).

2. Click on the small arrow to open the drop-down list and select **Save as New Style**. Self-descriptive names are always recommended but we are limited to 12 characters only for the style name. Type App Arch as the value and click **OK** to accept.

The name of your newly added style now appears in the styles drop-down list in the same menu items group. You can now apply this style to any application architecture element that you create simply by highlighting the element and selecting this style from the menu.

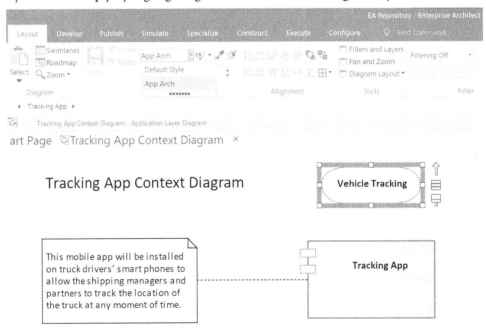

Figure 3.8 – The style of the highlighted element is saved for future reuse

Now we have the application service created and styled, we need to relate it to the Tracking App application component.

Relating the component to the service

According to the ArchiMate® 3.1 specification, the relationship between an application component and an application service is **assignment**. Assignment simply means that we are describing part of the Tracking App behavior in the Vehicle Tracking service. This relationship is represented as an arrow with a solid line and a dot at its base, as you can see in the following diagram:

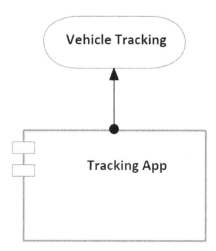

Figure 3.9 – The assignment relationship

We will talk in more detail about the structure and behavior of enterprise elements, along with more elaboration of the definitions of the relationship, in the next chapter.

We need to create an assignment relationship between the Application Component and Application Service elements, so we need to do the following:

1. Locate the assignment relationship in the **Toolbox,** click on it once, and you will see the shape of the mouse pointer changing to a hand symbol, indicating that you are now in relationship creation mode.

2. Click on the application component and keep holding the mouse button, move your mouse over the application service, and release it.

    Alternatively, you can click on the application component, click and hold the arrow-shaped action button, release it on the application service, and select **Assigned to** from the menu.

3. Press *Ctrl + S* to save.

This will create the desired relationship between the two elements. This relationship is not only visible on this diagram, but it also connects them internally and will appear automatically whenever you place both elements on any other diagram. This relationship is part of the enterprise architecture repository and is not limited to the current diagram.

To know the relationships between any element and other elements, highlight it and open the **Properties** pop-up window by pressing *Alt + Enter* or right-clicking on the element and choosing **Properties** > **Properties** > **Related** > **Links**, as shown in the following figure:

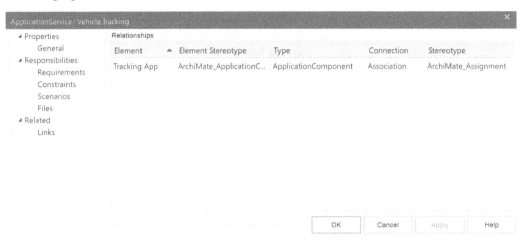

Figure 3.10 – Relationships between the current element and other elements

If you discover a need for adding additional application services, you can follow the same steps to add them. For now, your diagram must look as in the following figure:

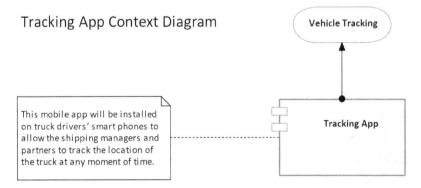

Figure 3.11 – Tracking App and the service it provides

As you can see, the diagram is starting to take shape, but we still need to add more to the context diagram, so let's keep adding more content.

Adding Actors

Now we need to place the users of the application on the diagram, and we will use the **Business Actor** element for this purpose. Business Actors are not part of the Application toolbox, so we need to bring in the ArchiMate® 3.1 Business toolbox instead, as we will see in this section. We will perform the following steps to add Business Actors to the application context diagram:

1.   Activate the ArchiMate® 3.1 Business toolbox.
2.   Place the Business Actors on the diagram.
3.   Define the proper relationships between the Business Actors and other elements.
4.   Style the Business Actors.

As you can see, the steps are similar to those we have followed to add the application service, but first, we need to activate the Business toolbox to be able to use the business elements.

Activating the ArchiMate® 3.1 Business toolbox

Sparx made ArchiMate® 3.1 Application the default toolbox for our diagram because we earlier chose that we wanted to create an ArchiMate® 3.1 Application diagram. However, in many cases, we may need to add elements from different toolboxes, so Sparx provides a way to do this. If you look at the top part of the **Toolbox**, you will see a *search box*, a *search button*, and a toolbox menu (also known as a *hamburger menu*), as shown in the following figure:

Figure 3.12 – The Toolbox menu items

The search box allows you to search within the active toolbox, the search button allows you to search within all Sparx toolboxes, and the toolbox menu allows you to change the toolbox and the **Perspective**.

Sparx groups toolboxes that serve a similar purpose in what it calls a Perspective, so there is a Perspective for ArchiMate® 3.1 that contains all the toolboxes under the ArchiMate® 3.1 specification, there is a Perspective for UML 2.5, a Perspective for Strategy, and so on. Use the toolbox menu whenever you need to use an element from a different toolbox within the same Perspective, or to change to another Perspective altogether.

> Note
>
> The toolbox menu uses a symbol that is *informally* known as the **hamburger menu** because of the way it looks. This is not standard naming, but it is well known among developers and most internet users.

Keep in mind that based on your purchased Sparx license, some Perspectives may show in the list, but they will contain no toolboxes, which means they are not included in your license.

To activate the Business toolbox, click the toolbox menu > **ArchiMate 3.1** > **Business**, where you can see all the business architecture elements.

Placing the Business Actors on the diagram

While reading the Use Case documentation, you will have realized that three main Business Actors will be using Tracking App: truck drivers, shipping managers, and business partners.

> Note
>
> The techniques for identifying Actors for a specific application fall more into the **business analysis** domain, which is not our focus in this book. We are assuming that some form of documentation already exists, and it does not have to be formal, but may exist in presentations, spreadsheets, or web pages.

To add the Business Actors to the diagram, make sure you do the following:

1. With the Business toolbox open, drag three **Business Actor** elements from the **Toolbox**, and drop them onto the diagram above the application service.
2. Rename the three Actors as Truck Driver, Shipping Manager, and Business Partner.
3. Add a short description for each Actor in the designated **Notes** field.

With more elements added to the diagram, you may realize that you need to move elements to a different location to make space for more elements to be added. Let's look at how to move the elements on a diagram.

If you need some space between the diagram label and the elements that you have placed, you can follow the same selection techniques that you are already familiar with in other tools, such as the following:

- You can *click and hold* your left mouse button to draw a selection rectangle. Every element that falls fully or partially within the selection rectangle will be selected.

- You can hold the *Ctrl* key on the keyboard and *click* on elements to select or unselect them.

- You can press *Ctrl + A* to select all elements on the diagram and then use the *Ctrl + click* technique to unselect the unwanted ones.

All these techniques are pretty standard and available in every modeling tool (and even in Windows File Explorer), possibly with slight variations, so there is nothing new to say here except to remind you that the same old techniques that you may know already work in Sparx too.

After selecting the elements that you want to move, click and hold any of them, drag them to where you want to move them, and release. Alternatively, you can move a group of selected elements by one point at a time using *Shift* + arrow keys to move the selection to the left, right, up, or down accordingly. This is a very helpful technique to slowly move objects for more accurate positioning.

Next, we will see how to relate the three Business Actors to the existing elements on the diagram.

Defining the relationships to Actors

It is important to keep in mind that everything is connected in the enterprise. There should be no single element in the entire enterprise repository that exists by itself and has no connections to at least one other element. Defining the proper relationships will help us to identify and analyze the impact of making changes, as we will see later in *Chapter 8, Business Architecture Models*.

The relationship between an application service and a Business Actor is of the **serving** type, and I will show you in the next section of this chapter how I know that. The serving relationship looks like a straight arrow with a solid line. The arrow base is connected to the element that is providing the service, and the arrowhead points to the element that receives the service.

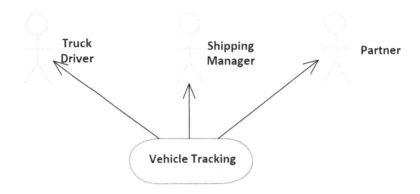

Figure 3.13 – An application service serving Business Actors

Since you know how the relationship looks, it will be easy to find it in the **Toolbox** and create it between the elements:

1. Locate the serving relationship in the **Toolbox**, click and hold the application service, move to the first Actor, and release.

2. Repeat the same for the other two Actors.

3. Another way to connect is to click on the source element (the application service) to highlight it, use the small action arrow that appears to the side, hold it, and release it above the target element (the Actor), and then select **Serving** from the context menu.

Just like we did with other elements that we have placed on the diagram, we need to style the Business Actors.

Styling the Business Actors

For styling business architecture elements such as Business Actors, we will use the same font size that we have used for the application elements. However, we will use a different color for the border to visually differentiate elements that belong to different architecture layers. After styling one Business Actor element, we will save the style and apply it to the remaining Business Actors, as you will see in the following steps:

1. Select one of the Business Actors and adjust its font size to **10**, set the fill color to *white* (RGB 255, 255, 255), and set the border color to *orange* (RGB 255, 165, 0).

2. Save this style for future reuse as `Biz Arch` so you can apply it to other business architecture-layer elements without repeating the styling steps every time.

    Now we have the style saved, we can apply it to any other element. Reusing styles is a very important factor in maintaining the consistency of diagrams within a single repository, especially when there are multiple users contributing to the content.

3. Highlight the other Business Actors on the diagram either together as a group or one by one and apply the saved **Biz Arch** style to them.

4. Optionally, you can change the notation of the Business Actors to the borderless notation.

5. Save your work by pressing *Ctrl + S*.

If you have forgotten how to perform any of the listed steps, I encourage you to refer to previous sections in this chapter where we have provided detailed step-by-step instructions for both the Application Component and Application Service elements.

If you have followed the steps so far, your diagram will look like the following:

Tracking App Context Diagram

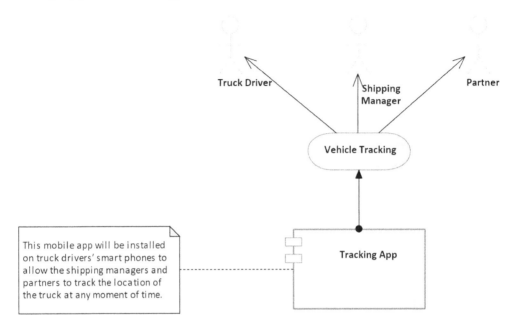

Figure 3.14 – Tracking App context diagram so far

The diagram is shaping up and is telling us more about Tracking App now. The project stakeholders may agree or disagree with what you are proposing but this is the main objective of conceptual diagrams. Our goal is to have all stakeholders on the same page of understanding, and opening doors for questions and answers always helps with this goal.

We still have more information to add to the diagram and one of them is the dependencies on other applications, so let's add it.

Identifying other dependent elements

With the identified services of Tracking App, there is not much information to exchange with other applications. Tracking App needs to send the phone's current location at specified time periods to the Trading Web application.

The Trading Web application is the main business application that internal users and external partners use to make orders, get item information, make payments, and access many other core online services. Identifying all these services will add great value to your enterprise repository, because the Trading Web application represents the information technology backbone of the ABC Trading business, and the more knowledge we can build about it, the more valuable information we can give to the decision-makers. However, this effort will shift your focus away from what you are tasked to do now. The best thing you can do in this case is to add a new user story to the artifacts backlog and discuss it with the product owner. For now, all we know is that our enterprise has an application component called Trading Web and it will exchange data with Tracking App.

Adding the Trading Web application component

We know that we have not created the Trading Web application component before, so we will create a new application component from the **Toolbox**:

1. Drag an application component from the **Toolbox** and drop it on the diagram to the side of **Tracking App**.

2. Rename it to Trading Web, style it by using the **App Arch** style, and add a description in the **Notes** area as desired.

3. Add descriptive text to the **Notes** field of the Trading Web component.

> Note
>
> Human Actors and other applications (which are technically considered Actors too) can use the same service if they need to access the same data. They may require two different types of interfaces to do so but it is still the same service and the same data. Human Actors (or users) require having a UI, while applications require an API to access the same service. We will talk in more detail about application services and application interfaces in the next chapter.

Tracking App will send the location of the truck (or vehicle in general) to the Trading Web app through the Vehicle Tracking service at every specified period. In other words, *Tracking App* will be *serving Trading Web* with *vehicle location data*.

4. Use the Toolbox or the arrow action button to create a serving relationship going from the Vehicle Tracking service to the Trading Web application component.

Next, we will style the Trading Web application component just like we did with the other elements.

Styling the Trading Web application component

Trading Web is an application component, so you can reuse the previously saved **App Arch** style as we have done earlier, or use this quick copy/paste method to apply a style from one element to another element on the spot:

1. Highlight the **Vehicle Tracking** application service.

2. Locate the **Layout** > **Style** > **Pickup** toolbar button, which has an *eye drop* icon. Click it, and the style will be copied to the clipboard and will remain there until you copy another style or close Sparx.

3. Now, highlight **Trading Web** and click the **Layout** > **Style** > **Apply** button – which has a *brush* icon – to apply the style that is in memory on the selected element.

4. Either keep the default rectangular notation or use the borderless one.

So far, we have learned two ways for reusing styles, and each has its own benefits:

- **Saving the style**: This is useful if you will use the same style over and over in different places within the repository for a long period of time, so it is more for *permanent* reusability.

- **Copy/paste the style**: This is useful for quickly applying a style of one element to another, and if you want it to stay in memory for a short period of time, so it is more for *temporary* reusability.

Save a style only if you are sure that you will reuse it later because you do not want to *pollute* the styles list with unused elements. If you are not sure that you will reuse the style again, it is more efficient and cleaner for the repository to copy/paste the style from another styled element.

> **Note**
>
> If you are working in a corporate environment and your repository is hosted in a centralized server, saved styles will be shared across all users. Therefore, only save styles that you will be reusing as a team and avoid saving styles that are just your personal preference.

This is what the diagram will look like at this point:

Tracking App Context Diagram

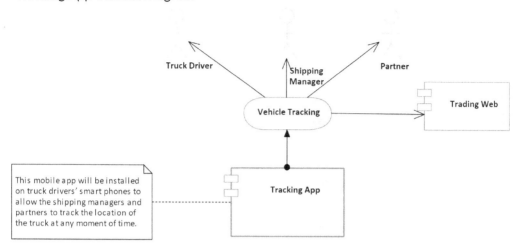

Figure 3.15 – The Tracking App context diagram up to this point

The last question that we need to answer is what technology services will be required to support this application, so let's do that.

Adding the supporting technology services

The location of the truck will be determined by the location of the phone that runs the application. This is a service that is already provided by the phone's operating system, so all we need to do is to use it. At this level of abstraction, we only need to mention that we will use the location service provided by the operating system. This type of service is known as **technology services**, which we will talk about in more detail in *Chapter 6, Modeling in the Technology Layer*, and *Chapter 7, Enterprise-Level Technology Architecture Models*.

Technology services are like application services since they both describe the exposed behavior of a structure element. Technology services are *business neutral* while application services have some *business logic* that is related to the business domain of the organization. The Vehicle Tracking service, for example, is a service that is of use only to a business that has requirements to track vehicles, so it is classified as an application service. The Device Location service, on the other hand, is a service provided by the phone regardless of what applications use it. It can be used by a navigation application, a marketing application, a social application, or even a game, so it is business neutral and therefore classified as a technology service.

We need to add a new Technology Service element to the diagram, but since the currently active toolbox is the Application architecture toolbox, we need to change it to the **Technology** architecture toolbox. Follow these steps to add a new Technology Service element to the diagram:

1. Click on the hamburger action menu, and then select the **ArchiMate 3.1 > Technology** architecture. This will activate the ArchiMate® Technology architecture toolbox.

2. Locate the Technology Service element in the Toolbox, drag it, and drop it on the diagram below the Application Component element.

3. Rename the service to `Device Location`, change the font size to **10**, change the fill color to *white* (RGB 255, 255, 255), and change the border color to *dark green* (RGB 0, 100, 0).

4. Save the style as a new `Tech Arch` style for future reusability.

5. Add descriptive text to the **Notes** field describing the technology service.

6. Use the borderless notation if you prefer, or keep the default rectangular notation.

7. The Device Location service will provide *location data* to the application component. In other words, the *Device Location service* is *serving* the *Tracking App component* with the *phone's location data*. Therefore, create a relationship of the *serving* type going from the technology service to the application component.

8. Save the changes by pressing *Ctrl + S*.

Showing relationships between two elements is good and showing what gets transferred through these relationships is even better. We will see next how to model the data that goes from one element to another.

Adding the sent data

The operating system will provide the device's location in the form of data. We do not know a lot about the structure of the data at this time, but we know that data will be provided from the Device Location service to the Tracking App component, so we need to model that. There are multiple ways to model how data flows from one element to another. I will use one way for this case and will show you the other ways in other examples as we progress:

1. Open the **ArchiMate 3.1 > Application** toolbox and locate the **Data Object** element.

2. Drag the **Data Object** element and drop it onto the diagram near the **Device Location** service.

3. Change the data object's name to `Device Location Data` and style it as **App Arch**.

4. Add a proper description to the **Notes** field describing the data object.

    Notice that there is only one notation for data objects so you will not see the option to change it in the context menu. Now we need to indicate that the data object will be provided or *served* by the Device Location technology service to the Tracking App component.

5. Click on the serving relationship and you will see a small action arrow that appears to the right of the relationship.

6. Hold this arrow, drag it to Device Location Data, and select **Associated with**. This creates an association relationship between the serving relationship and the data object and tells the user what data will be served.

7. Press *Ctrl* + *S* to save.

If you were following properly, then your diagram will look similar to the following one:

Tracking App Context Diagram

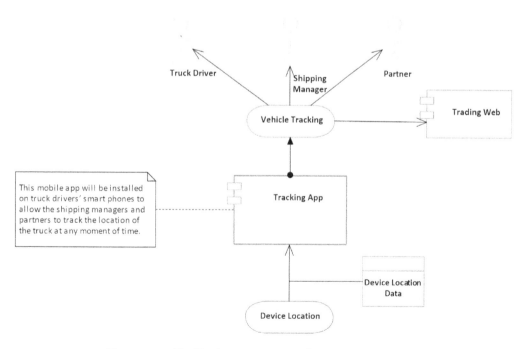

Figure 3.16 – The Tracking App context diagram in its final state

The diagram is mature enough at this point to be shared with stakeholders.

If a stakeholder wants to deliver a presentation on the application, having this diagram can answer a lot of the questions in a very simple yet well-documented way. The diagram gives enough information at a conceptual level; you must keep your context diagrams at this level of abstraction.

The application context diagram is meant to build a common understanding of a specific application component. By looking at it, any user at any level of experience will have the same understanding of what the application component is or is supposed to be. There is no technical jargon, no abbreviations, no third-party products, and no specific implementation enforcement. It is meant to have every stakeholder on the same page.

In the next chapter, we will create a simple and easy-to-read metamodel, which will help us when we create more diagrams that have the application component as the main element.

Summary

This chapter has introduced many new concepts that will be followed throughout the remaining chapters of this book. We have learned how to create packages in Sparx, how packages can contain diagrams and elements, and how to build nice-looking yet useful diagrams. EA artifacts in general, and diagrams in particular, are what architects use the most to communicate information. Mastering the skills of modeling is a key to your success as a practical enterprise architect.

In the next chapter, we will look at how to navigate the ArchiMate® 3.1 standard and how to build your own standard reference, the focused metamodel. This is a very important aspect of building and maintaining trust in your enterprise repository, so keep moving forward before losing momentum.

4
Maintaining Quality and Consistency in the Repository

After going through *Chapter 3*, *Kick-Starting Your Enterprise Architecture Repository*, you now have an enterprise architecture repository with a single artifact, an application context diagram for the **Tracking App** package. At this point, we need to step back and imagine where all of this may lead. This is just the first of many artifacts that are needed to represent your model-driven enterprise. You might be wondering how you can scale this effort so that others in your organization can help. You need a standard against which you can measure the quality and consistency of artifacts from various authors. Without such a standard, you would quickly run into problems with consistency and quality in your repository. Asking everyone in your organization to become experts in the ArchiMate® 3.1 notation is unreasonable.

In this chapter, we delve into modeling best practices and how to represent your model standards using the concept of a **metamodel**. Just as metadata is data that describes other data, a metamodel is a model that describes other models. Here, we explore the ArchiMate® standard metamodels and learn to condense that information into what we call a **focused metamodel**. These are the topics that will be covered in this chapter:

- Building the application component-focused metamodel
- Modeling best practices

But before we start, let's make sure that you have the necessary technical requirements

Technical requirements

We will be referring to the *ArchiMate® 3.1* standard in this chapter, so access to The Open Group®'s ArchiMate® reference (`https://pubs.opengroup.org/architecture/archimate3-doc/`) is important. It would also be helpful, but not necessary, to have some familiarity with that work.

Building the application component focused metamodel

In *Chapter 1*, *Enterprise Architecture and Its Practicality*, we talked about **metamodels** in general and **focused metamodels** specifically. We explained that metamodels are the blueprints that guide the development of architectural artifacts such as diagrams. We developed our first diagram in the previous chapter without a metamodel, but that is because we were referring to the ArchiMate® 3.1 metamodels without telling you, to avoid confusing you at the time. But now, since you have more experience in developing diagrams, we feel more confident talking about metamodels and will show you how to create focused metamodels that are easier to understand and follow than the standard ones.

Focused metamodels are our interpretation of ArchiMate®'s standard metamodels. ArchiMate® metamodels are great references, but they are organized by layer. To get the full picture of an element and to find all possible relationships across the different layers, you will need to go back and forth between multiple diagrams in different chapters of the ArchiMate® specification, as you will see shortly. Focused metamodels combine all that you need to know about a specific element in one place, so they are *element-oriented*, not *layer-oriented*.

Each focused metamodel has one element in focus at a time and shows the possible relationships to other elements. You can enrich your enterprise architecture repository with dozens of focused metamodels, each focused on a different element. On the other hand, you can still work without them and use the standard metamodels, but they will help make your enterprise architecture practice easier. Remember that when you work in a team, each member can have a different level of experience in ArchiMate® and enterprise architecture, so having an easy-to-follow reference can be very handy to unify team efforts.

Metamodels are diagrams that guide us in making diagrams. To create a metamodel, we will follow the same steps that we took to create the application component context diagram in the previous chapter. The only difference between the two diagrams will be the content. Sparx treats both diagrams the same and does not provide any special treatment to one diagram over the other. It is your responsibility as an enterprise architect to organize the repository in a way that makes sense to you and the audience.

In this section, we will create our first focused metamodel and it will be the application component focused metamodel. We will achieve this by doing the following:

- Establishing the metamodel diagram by creating a package and a diagram
- Using ArchiMate® 3.1 as a reference to guide the focused metamodel creation
- Adding the proper elements to the diagram according to the reference material

Since you have some familiarity with the process already, we will not go through the same level of detail that we went through in the previous chapter, so you may find yourself needing to go back and check some instructions there if you are not sure—therefore, keep this in mind.

Establishing the metamodel diagram

The first thing that we need to start with is to create a new package that will contain the new diagram and the elements that will be placed on it. In other words, we will be doing the following in this subsection:

- Creating a **Metamodels** package and the application component-focused metamodel diagram
- Adding a focus element to the diagram

Remember that we have already created a root node package and named it **Architecture Content**. We will create a new **Metamodels** package under the root node package but at the same level as the **Tracking App** package that we created in the previous chapter. Metamodels are references for architects working on any enterprise architecture artifact for any element within the enterprise. On the other hand, the **Tracking App** package contains the **Tracking App Context Diagram** and the elements that are specific to the diagram. Therefore, we need to place the **Metamodels** package under the **Architecture Content** package instead of under the **Tracking App** package.

Creating a diagram in a package

We will follow the same method that we followed to create the context diagram by creating a package with a diagram and then adding elements to it, as follows:

1. Click on the **Architecture Content** node package, and then click on **Design** > **Model** > **Add** > **Package**, which will bring up the **New Package** pop-up window. Type Metamodels in the **Name** field, select **Create Diagram**, and click **OK**.

2. The **New Diagram** pop-up window will appear automatically, so enter the name Application Component Focused Metamodel in the **Diagram** field.

3. From the **Type** list, either choose **All Perspectives** to list all diagram types or choose **Enterprise Architecture** > **ArchiMate 3.1** to filter the list to ArchiMate® 3.1 diagrams only.

4. Click on **ArchiMate 3.1**, and a list of available ArchiMate® diagrams will be shown in the right section of the window under **Diagram Types**.

5. Choose **Application** because we are creating an application-type diagram.

6. Click **OK** to confirm your choices.

    You will see that a new **Metamodels** package has been created with a single empty **Application Component Focused Metamodel** diagram in it. Next, we need to add elements to the diagram.

    In the previous chapter, we used a text element as a diagram label, which gave us the flexibility to change the font size, the location of the label, and the color, and we are free to type anything we want to type. This sounds great but it comes with a small cost, which is maintaining its content. If you rename your diagram (right-click on the diagram, then **Properties** > **General** > **Name**), you need to rename the label content as well to reflect the new name, or else your content will be outdated.

    Sparx has provided another way to add labels to diagrams, which has different advantages and disadvantages. Let's add a diagram label next, and we will discuss its advantages and disadvantages after that.

7.   Right-click on the diagram, and then select **Properties** > **Diagram** > **Appearance** >
     Show Diagram Details.

8.   Click **OK** to accept the changes.

Notice that the diagram now shows four lines of information in the top-left corner: **Name**,
Package, **Version**, and **Author**. The best advantage is that when you rename the diagram
or package that contains this diagram, the label will update automatically to reflect the
new names of the diagram or package. However, the biggest disadvantage is that you have
no control over what to show or what to hide in this label, no control over the font size or
color, and no control over the location. The label will always be displayed in the top-left
corner, with the same four lines of information, the same font size, and the same font
color. You either accept it as it is or reject it completely as there is no current possibility to
customize it (at least up to version 15.2 of Sparx, which we are currently using).

You have two ways to choose how to display diagram labels, so it is up to you to follow
either of them. Just make sure that you remain consistent with your choice. The diagram
still has no elements on it, so let's start by adding the focus (the primary) element.

Adding a focus element to the diagram

The focus element in our case is the **Application Component** element. This is the element
that we are creating a diagram about, so it will be visually differentiated from other
elements on the diagram by the following styling characteristics:

*   It will be in the center of the diagram.

*   It will be relatively larger than the other elements.

*   Its border will be thicker than the borders of the other elements.

Let's add the application component to the diagram and make it the focus element. If
you do not have the application component metamodel diagram opened, please do so by
double-clicking it from the **Project Browser** and then go through the following steps:

1.   From the **Toolbox**, locate the **Application Component** element and drag and drop
     it to the center of the diagram area.

2.   Click on the **Application Component** element either on the diagram or in
     the **Project Browser**, and then press *F2* on the keyboard to rename it from
     `ApplicationComponent1` to `Application Component`.

3.   Add the definition of **Application Component** in the **Notes** area, and use the
     complete definition from ArchiMate® 3.1 specification.

4. Style the application component as **App Arch**, change the thickness of the border to **2**, and, optionally, use the borderless notation or the rectangular notation.

5. Press *Ctrl + S* to save.

We have kept the rectangular notation for the application component in this diagram while using the borderless notation for the other elements. This is a styling choice, and we will be using the rectangular notation for all focus elements within the metamodels. Any style is a good style if you remain consistent in using it.

Interpreting an ArchiMate® metamodel

We need to add elements that can relate an element of the application component type to the diagram. We are following the ArchiMate® 3.1 specification in this book, so will use the online material as a reference to derive our easier-to-read metamodels. The complete reference is available at `https://pubs.opengroup.org/architecture/archimate3-doc/toc.html`, but we will mainly use the following ArchiMate® 3.1 chapters in this book:

- *Chapter 5, Relationships* (`https://pubs.opengroup.org/architecture/archimate3-doc/chap05.html#_Toc10045310`)

- *Chapter 8, Business Layer* (`https://pubs.opengroup.org/architecture/archimate3-doc/chap08.html#_Toc10045365`)

- *Chapter 9, Application Layer* (`https://pubs.opengroup.org/architecture/archimate3-doc/chap09.html#_Toc10045389`)

- *Chapter 10, Technology Layer* (`https://pubs.opengroup.org/architecture/archimate3-doc/chap10.html#_Toc10045407`)

- *Chapter 12, Relationships Between Core Layers* (`https://pubs.opengroup.org/architecture/archimate3-doc/chap12.html#_Toc10045440`)

For copyright reasons, we cannot include the diagrams from the online material in our book, but you can bookmark the provided references in your browser, download them to your computer, or print them if you like—whichever you feel is more convenient for you.

Looking carefully at ArchiMate®'s metamodel, you can see that it has a lot of useful information, but it can be slightly difficult to read because of the use of generic type names such as **Application Internal Active Structural Element** to refer to an application component.

In this subsection, we will show you how to interpret the standard ArchiMate® 3.1 metamodels in step-by-step instructions, to build an easier and more focused metamodel that can be used as a future reference. We will do the following:

- Start by analyzing the reference metamodels and understand how to read them.

- Extract the statements that are conveyed within the metamodels and use them as building blocks for our focused metamodel.

- Build our first focused metamodel.

The more you get familiar with the ArchiMate® specification, the easier it becomes to interpret its metamodels, so let's start developing this skill.

Analyzing the reference metamodels

Look carefully at *Figure 70: Application Layer Metamodel* in the *ArchiMate® 3.1 Specification, Chapter 9* (`https://pubs.opengroup.org/architecture/ archimate3-doc/chap09.html#_Toc10045390`) and notice the following:

- There is one element called **Application Internal Active Structure Element**. This, according to the specification, can include the **Application Component** and **Application Collaboration** elements.

  Application collaborations are simply groups of application components collaborating to fulfill a given objective. An **enterprise resource planning** (**ERP**) solution, for example, is an application collaboration as it can aggregate multiple different application components, but they all serve the same objectives. Structure elements describe *what* makes up the application components.

- The **Application Interface** element is also a structural element, but it is the part of the application that is exposed to the enterprise , so it is considered as an **Application External Structure Element**.

- The **Application Internal Behavior Element** includes **Application Process**, **Application Function**, and **Application Interaction**.

  All these elements describe the behavior of a component, such as—for example— transforming data from one format to another. Behavior elements describe the *how* part of the application components.

- The **Application Event** and **Application Service** elements also describe behavior, but they are exposed to the external enterprise, so they are **Application External Behavior Elements**.

- Finally, there is the **Data Object** element, which is a **Passive Structural Element**. This means that it cannot perform any actions on other elements, but other elements can *access* it and perform actions on it.

We've just introduced lots of new terminologies, which might make your brain dizzy for a minute, so do not worry about understanding all of them at once because we will introduce them with complete definitions and supportive examples as we progress in this book. All that we need you to do now is to look at the **Application Internal Active Structure** element because this element **generalizes** the application component that we will use for our focused metamodel.

Identifying general statements

The next step is to identify the relationships that exist between the **Application Internal Active Structure** element and the other elements and put them into statements. We can identify the following generalized relationships from the diagram:

- An **Application Internal Active Structure** element can be *assigned to* an **Application Internal Behavior** element.

- An **Application Internal Active Structure** element can be *assigned to* an **application event**.

- An **Application Internal Active Structure** element can be *composed of* **application interfaces**.

- By default, an **Application Internal Active Structure** element can be *self-composed*, so it can be composed of or decomposed into other **Application Internal Active Structure** elements.

- Finally, an **Application Internal Active Structure** element can be *served by* an **application service**.

All of the mentioned relationships are defined in the referenced ArchiMate® metamodel diagram. If you cannot find them, please print a copy of that diagram on paper, put it side by side with the preceding list, take as much time as you need, and make sure that you can see and find all of them.

Building the focused metamodel

Let's take a moment to analyze the diagram and use the application component as a specialized element of the **Application Internal Active Structure** element. We can rephrase the first statement into three specialized statements, as follows:

- An **application component** can be *assigned to* an **application process**.
- An **application component** can be *assigned to* an **application function**.
- An **application component** can be *assigned to* an **application interaction**.

Let's have a quick overview of the formal definition of the *assignment* relationship before we go any further.

The assignment relationship

"The assignment relationship represents the allocation of responsibility, performance of behavior, storage, or execution." (`https://pubs.opengroup.org/architecture/archimate3-doc/chap05.html#_Toc10045314`)

You can see a visual representation of this relationship here:

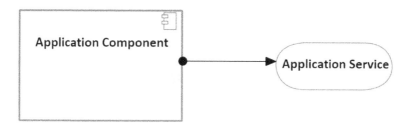

Figure 4.1 – The assignment relationship

This means that part of the application component's behavior is defined within elements that it is assigned to, such as an application service. If an application component performs transactions such as processing claims, *how* claims are processed is defined in the form of claim processing functions, processes, services, and interactions, which are all behavioral elements. As a general rule, structural elements are assigned to the behavioral elements, because you need to have a structure in order to perform the behavior. You cannot perform the behavior from nothing.

Application events also describe behavior that affects an application component; a very good example of application events are batch jobs that run at specific times. Now, we have a fourth assignment relationship statement, as outlined here:

- An **application component** can be *assigned* to an **application event**.

Before our list of bulleted items grows too long, let's add the identified ones to the diagram. As we mentioned earlier in this section, the behavior of an application component can be described using different behavioral elements such as application processes, functions, interactions, and events.

We need to add these elements to the diagram and define the possible relationships that can be created between the application component and these behavioral elements, as follows:

1. From the **Toolbox**, drag a new **Application Process** element and drop it onto the diagram area.

2. Rename the newly created element from `ApplicationProcess1` to `Application Process`.

3. Style the application process in **App Arch** style.

4. Select the desired modeling notation, whether rectangular or borderless.

5. Find the *assignment* relationship in the **Toolbox** and create an assignment relationship from the application component to the application process.

6. Repeat all previous steps to add **Application Function**, **Application Interaction**, and **Application Event** elements to the diagram, and don't forget to create an assignment relationship too.

7. Press *Ctrl + S* to save.

The diagram should look similar to the one shown here; you can, of course, organize the elements in a different order if you like, so you do not have to place them in the same locations that we did:

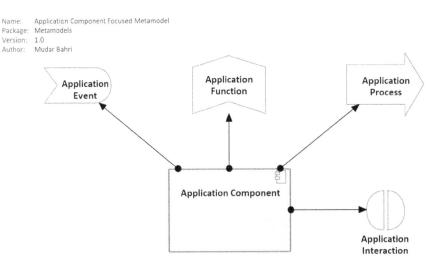

Figure 4.2 – The application component focused metamodel so far

Notice that we are still adding elements, so we may need to move some of the elements later to make room for additional ones.

The composition relationship

The third and fourth generalized statements that we have captured from interpreting ArchiMate®'s metamodel tell us about another type of relationship, which is the **composition** relationship.

"*The composition relationship represents that an element consists of one or more other concepts.*" (https://pubs.opengroup.org/architecture/archimate3-doc/chap05.html#_Toc10045312)

You can see a visual representation of this relationship here:

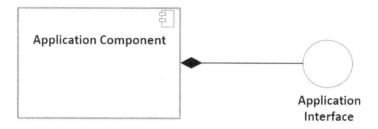

Figure 4.3 – The composition relationship

Composition is the well-known *parent-to-child* or *whole-to-part* relationship. When an element, A, is composed of elements B, C, and D, it means that element A is the whole, and elements B, C, and D are its parts.

An interesting thing to know about the composition relationship is that it is **transitive**. This means that if a composed element is related to another element, it implies that the parent composing component also has the same relationship with that element. For more information about deriving relationships, please refer to ArchiMate®'s online documentation at https://pubs.opengroup.org/architecture/archimate3-doc/apdxb.html.

Another interesting thing about the composition relationship is that any element can be composed or decomposed of larger or smaller elements of its type. So, an application component can be composed of many smaller application components, an application process can be composed of many smaller application processes, and so on.

Look back at ArchiMate®'s metamodel and we will explain how these relationships apply to the application component. First, there is a clear composition relationship between the application component and the application interface. There is also an implied self-composition relationship between the application component and itself. Let's write these down before we forget them, as follows:

- An **application component** is *composed* of **application interfaces**.

- An **application component** is *self-composed* of **application components**.

Additionally, there is an assignment relationship between the application interface and the application service. Since the composition relationship is transitive, it also means that the same relationship exists between the application component and the application service. Therefore, we can add the following statement to our list:

- An **application component** is *assigned to* an **application service**.

We now have sufficient elements and relationships to add to the diagram. We need to add the **Application Interface** element and add the proper relationships as we have stated them, so continue with the following steps:

1. From the **Toolbox**, drag a new **Application Interface** element and drop it onto the diagram area.

2. Rename it `Application Interface`.

3. Style it in the **App Arch** style.

4. Select the desired modeling notation, whether rectangular or borderless.

5. Find the *composition* relationship in the **Toolbox** and create a relationship from the application component to the application interface.

    The black diamond-shaped end of the relationship must be connected to the parent element, which is the application component.

6. Now, you need to create a self-composition relationship for the application component, so click the *composition* relationship from the **Toolbox**, click on the application component, move the mouse a little to the top, and click once again on the application component.

7. Do not forget to add the **Application Service** element and create an assignment relationship from the application component to it.

8. Press *Ctrl + S* to save your work.

Your diagram should look close to the following:

Name: Application Component Focused Metamodel
Package: Metamodels
Version: 1.0
Author: Mudar Bahri

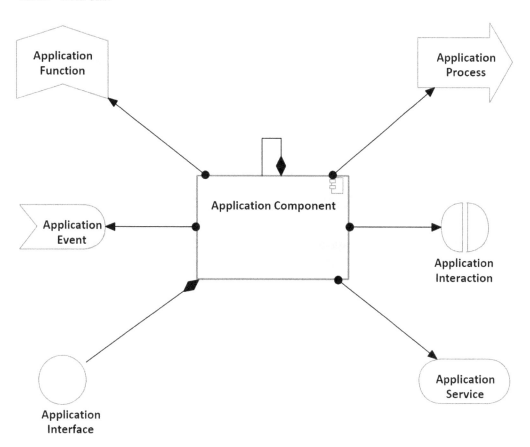

Figure 4.4 – The application component focused metamodel up to this point

Notice how we have reorganized the diagram to fit the newly added elements, but everything still looks well organized. This is a habit that we want you to develop while building diagrams—always keep them organized as much as you possibly can, even if you are still drafting them.

The aggregation relationship

Another form of the composition relationship is known as **aggregation**.

"The aggregation relationship represents that an element combines one or more other concepts." (https://pubs.opengroup.org/architecture/archimate3-doc/chap05.html#_Toc10045313)

You can see a visual representation of this relationship here:

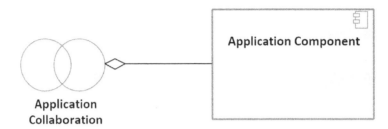

Figure 4.5 – The aggregation relationship

Aggregation is similar to composition as they are both two forms of the *parent-to-child* or *whole-to-part* relationship; however, aggregation is *weaker* than composition and the child can live completely independently from the parent. In composition, the child element is a *part* of the parent element and cannot function outside it, so it is more like containment. In aggregation, on the other hand, the parent element is a *combination* of its independent children. Each can work independently, and each can work combined with other elements under different parents.

To help you understand the difference between composition and aggregation, think of a house, for example. It is composed of bedrooms, a living room, and a kitchen. All these components are indivisible parts of the house, and they cannot exist by themselves. They must exist within a house to have a useful purpose, and the house is not a fully functional house without them. On the other hand, a detached shed or a detached garage are independent components, but when they are related to the house, they form a better set of options. You can disassemble the shed, move it, or sell it without affecting the integrity of the house, and the shed will be perfectly installed somewhere else.

Another example is an ERP solution that can be made up of accounting, inventory management, sales, billing, **human resources** (**HR**), marketing, and much more. Each of these can be a standalone solution and can work separately, but together, they are parts of a bigger solution. ArchiMate® calls the big solution that is a combination (aggregation) of multiple application components an **application collaboration**. Based on this, we can state the following relationship:

- An **application component** can be *aggregated in* an **application collaboration**.

Next, we need to add this element to the diagram. Following the steps that we took to add the other elements, we need to add an **Application Collaboration** element and connect it with an aggregation relationship to the application component, as follows:

1. Find the **Application Collaboration** element in the **Toolbox** and use it to add a new element to the diagram.

2. Rename the element `Application Collaboration`, style it with the **App Arch** style, and optionally change the notation to borderless.

3. Find the *aggregation* relationship in the **Toolbox** and create one from the **Application Collaboration** element to the **Application Component** elements. The diamond-shaped end must always be connected to the parent element.

The diagram now shows that an application collaboration is an aggregation of application components.

The access relationship

"The access relationship represents the ability of behavior and active structure elements to observe or act upon passive structure elements." (`https://pubs.opengroup.org/ architecture/archimate3-doc/chap05.html#_Toc10045319`)

You can see a visual representation of this relationship here:

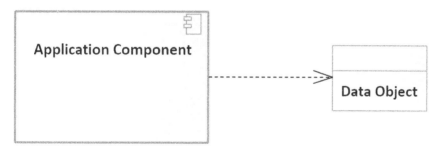

Figure 4.6 – The access relationship

In simpler English, the access relationship means that an element (whether structural or behavioral) can access the indicated data object. Based on the ArchiMate® application metamodel, we can see that an application component is assigned to an application behavior element, which accesses the data object. Therefore, we can say the following:

- An **application component** *accesses* a **data object**.

Now, we need to add a **Data Object** element to the diagram and connect it with an access relationship, as follows:

1. Drag a **Data Object** element from the **Toolbox** and drop it onto the diagram area.

2. Rename it Data Object, and style it as **App Arch**.

    You cannot change the notation of data objects, so there is no possibility to choose between the rectangular and borderless notations.

3. Find the *access* relationship in the **Toolbox** and create one from the application component to the data object.

Keep in mind that an access relationship can *only* exist between active elements on one side and passive structure elements on the other side. It cannot exist between two active structures, two active behaviors, or even between an active structure and an active behavior element. The arrowhead of the relationship must always be connected to the data object.

The serving relationship

The last relationship that we have identified is **serving**.

"The serving relationship represents that an element provides its functionality to another element." (https://pubs.opengroup.org/architecture/archimate3-doc/chap05.html#_Toc10045318)

You can see a visual representation of this relationship here:

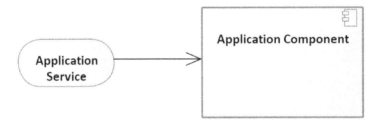

Figure 4.7 – The serving relationship

If element A is serving element B, it means that element B is using the functionality that is provided by element A. It can be read in the opposite direction, as element B is served by element A. A web application is served by a web server, for example.

As per ArchiMate®'s metamodel, we can add the following two statements:

*   An **application component** is *served* by an **application service**.

*   An **application component** is *served* by an **application interface**.

We already have the **Application Component**, **Application Service**, and **Application Interface** elements on the diagram, so we only need to add the relationships, as follows:

1. Find the *serving* relationship in the **Toolbox** or use the arrow-shaped action menu item to create a serving relationship going from the application service to the application component.

2. Repeat the previous step to create a serving relationship going from the application interface to the application component.

Because there are relationships already in place between the component and the service and between the component and the interface, Sparx will place the newly added serving relationships above the old ones, so you will need to manually adjust and separate them. Click and hold the relationship, then drag one away from the other. This is what the diagram should look like at this point:

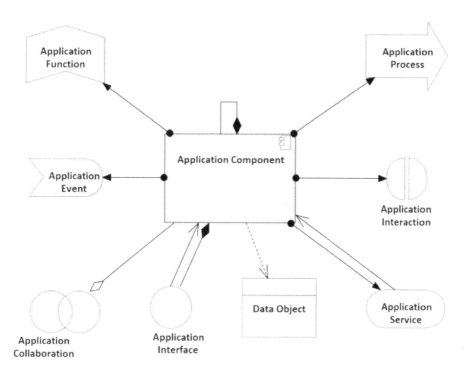

Figure 4.8 – The application component-focused metamodel up to this point

Don't forget to save your work before we continue. There are still more elements and relationships to add to the focused metamodel. They belong to different architecture layers, so we need to refer to additional reference materials to know what these elements are and how they are related to the **Application Component** element.

Adding elements from other architecture layers

All the elements that we have placed on the metamodel so far belong to the Application architecture layer, but the application component is also related to other elements from other layers such as the Business and Technology architecture layers. ArchiMate®'s metamodels do not show all the possible relationships in one place, and you will find yourself moving back and forth between multiple locations to get the full picture and find out all the possible relationships to an element. In this section, we will do the following:

- Find the elements from the Business and Technology architecture layers that can have a relationship with **Application Component** and identify these relationships.

- Add elements from the Business and Technology layers to the application component-focused metamodel diagram.

Once you see the completed focused metamodels, you will understand why they are handier and easier-to-read references than standard metamodels.

Finding and reading references

In ArchiMate® 3.1, the relationships between elements from different layers are detailed in *Chapter 12* of the specification (`https://pubs.opengroup.org/architecture/archimate3-doc/chap12.html#_Toc10045440`). As we mentioned earlier, the biggest advantage of focused metamodels is that they save you time searching for the complete picture in multiple chapters and diagrams within the ArchiMate® specification.

For now, we still need to use the specification until our focused metamodels become complete and reliable. We can see the following:

- An **Application Internal Active Structure** element (such as **Application Component**) can be served by **Business Service** and **Business Interface** elements.

- An **Application Internal Active Structure** element can be served by **Technology Service** and **Technology Interface** elements.

Now we know all this information, we need to add it to the focused metamodel to make it as complete and reliable as we can. Remember that we can always come back to these metamodels and add new elements when we need to, so it does not have to be 100% complete at this point and it is anti-agile to make it so.

Adding business and technology elements

Follow these steps to add business architecture elements to the diagram:

1. Change the **Toolbox** to the business architecture toolbox. You can do that by clicking the hamburger menu on the left corner of the **Toolbox** window and selecting **ArchiMate 3.1 > Business**.

2. Locate the **Business Interface** element, drag it, and drop it onto the diagram area.

3. Rename it `Business Interface`, style it as **Biz Arch**, and optionally change the notation to borderless.

4. Find the serving relationship in the **Toolbox** and create a connection from the business interface to the application component.

5. Repeat *Steps 2* to *4* to add a **Business Service** element and create a serving relationship to the application component.

6. Press *Ctrl + S* to save your work.

Repeat almost the same steps to add the technology architecture elements (**Technology Interface** and **Technology Service**). We will leave this as an exercise for you, but remember to change the **Perspective** setting to **ArchiMate 3.1 > Technology** and apply the **Tech Arch** style on the technology elements this time. Your diagram must look like this, or close to it:

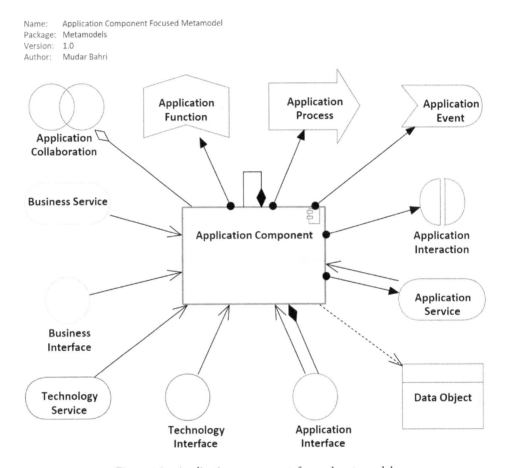

Figure 4.9 – Application component-focused metamodel

We have sufficient details on the focused metamodel to guide us in developing new application component diagrams. We have the definition of what an application component is (in the notes), we know most of the elements that can relate to it, and we know which types of relationships we can use. Having all the information that we need in one place is way easier than digging through the reference material every time we need to know something.

We do not need to use all these elements in every application component diagram, but we only use what makes sense to the audience per diagram. The application component context diagram that we developed in the previous chapter used some of the elements that are shown in the metamodel. Other application component diagrams may use a different set of elements based on what they are addressing. Say, for example, that you are developing a new application services catalog diagram. The only two elements that you need to use are the **Application Component** and **Application Service** elements.

The relationships between elements will remain the same as per the metamodel in every diagram. The relationship between an application component and an application service, for example, is always assignment, regardless of the type, scope, and audience of your diagram.

The next section provides some guidelines for making your diagrams eye-catching even if you want to follow different styles from the ones we used.

Modeling best practices

It is always a good practice to organize your diagrams in a way that makes them attractive to readers. Aligned elements on a diagram look better than misaligned ones, elements with similar spaces between them look better than elements with random spacing, and elements of the same size look better than elements of random sizes. So, do your best to have your diagrams as organized as possible because the difference this practice makes it worth the effort.

In this section, we will be giving some guidelines that we had set for us when creating diagrams. You can follow these guidelines as they are, or you may add to them from your own experience as well. These are not part of any standard but are based on our personal experience in modeling, so nothing is written in stone. You may agree or disagree with us on some, we may learn something new tomorrow that results in changing some of these guidelines, or you may have a different experience background than ours, so you may add your own touch and flavor. We all learn something new every day, no matter how old or experienced we are.

Keeping your diagram focused

A single diagram better tells a single main idea. If you put many ideas in a diagram, the readers will get confused about what are you trying to address. As you have seen in the application context diagram we have created, the focus was on the **Tracking App** component, and the entire diagram was made to serve the single idea of describing it. If we had added some unnecessary details about the **Trading Web** application component, for example, the diagram would be confusing to the reader as it would be describing two components at the same time.

You can see an example of such a diagram here:

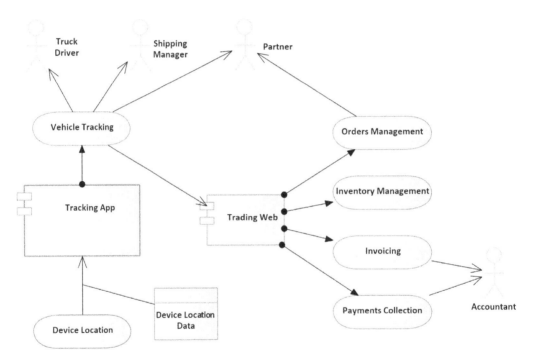

Figure 4.10 – Avoid introducing multiple ideas in a single diagram

As you can see in the preceding diagram, we have added some of the services that the **Trading Web** component provides, and added a new **Accountant** business Actor who will be using **Trading Web** but has nothing to do with **Tracking App**. This information is irrelevant to the **Tracking App** context diagram and only adds noise to it. Even though the preceding diagram is still correct, it has more information than the reader needs to know. Readers will be confused about whether you are telling them about **Tracking App**, **Trading Web**, or all the available application services.

You can tell readers more about the **Trading Web** component and who will use it, but this is better off in a separate diagram that has **Trading Web** as the focus and adds the other elements as context around it. Knowing how the **Accountant** Actor will be served by an application that is not **Tracking App** is unnecessary noise and will only cause confusion. On the other hand, it will be helpful to know how all Actors are being served and by which applications, but again, this is better off in a separate diagram that only focuses on Actors and services.

Sometimes, you may need to have many elements in one diagram, which is fine if the idea that you are trying to show in the diagram is how complex the situation is. If the purpose of your diagram is to list all the application components—for example, all the application services that exist within the enterprise, and all the data objects they exchange—then you will end up with a complex diagram with lots of overlapping connectors and shared data sources. But in this case, *complexity is the idea* that you are conveying, so avoid introducing a second idea such as where the application instances are deployed or which technology stacks each application requires.

Remember that in Sparx, you can create child diagrams for any element in the repository, which allows you to create drill-down diagrams showing different levels of detail. If your diagram looks complex and your purpose is not to show complexity, we highly advise you to break down the diagram into multiple smaller and simpler diagrams.

Fitting your diagram onto a single page

It may sound difficult, but if you follow the first advice of keeping your diagrams focused, this one will be easy to achieve. There is no harm in having large diagrams that span over multiple pages, but keep in mind that not every reader will be reading your diagrams from the Sparx **user interface** (**UI**), and not every reader has a monitor the same size as yours.

Additionally, if you publish your enterprise architecture content in a Word document, large diagrams will be shrunk automatically to fit within the page. This is acceptable if your diagram uses a little extra space beyond the page border as this will not affect its readability, but if it uses more than two pages, readers will find it extremely difficult to read it, and it will be of no value to them. It is hard to realize that such an effort that might have taken hours of your time is ultimately useless.

You can set the size of a diagram to any standard paper size or set a custom size if needed by right-clicking on the diagram and selecting the suitable page size from the **Properties** > **Diagram** > **Advanced** > **Page Setup** > **Paper Size** drop-down list. You can use the same dialog box to change the page orientation from **Portrait** to **Landscape** or the opposite. You can show/hide page borders by checking/unchecking **Properties** > **Diagram** > **Hide Page Border**. Notice that you can do this for a single diagram or all diagrams.

Adding only the necessary information

In some cases, showing details can be beneficial, and in other cases, hiding them can be more beneficial. It all depends on the purpose and the target audience. Take, for example, the **Tracking App** context diagram. We have shown that it uses a single technology service, which is the **Device Location** service, but this is not a complete list of services, to be honest. For an application to work on a device in an environment (whether it is a personal computer, a smartphone, a gaming console, or a **virtual machine** (**VM**)), it needs to be provided by many other services, such as storage, computing, network gateways, subnets, load balancers, and containers, to name but a few, as the list can go on much longer. Having all these services listed in a diagram that is supposed to provide an overview of what an application is can end up containing more details than what the target audience needs to know, and it needs to be avoided.

However, if our target audience is a cloud architect, for example, we will need to list all the cloud services that will be needed, but that will be another diagram: a separate diagram that targets a different audience and tells a different message.

Paying attention to your diagram's appearance

As an architect, most of your deliverables will be in the form of diagrams. The quality of your diagrams *will* reflect your quality as an architect, so make sure that you are reflecting the best possible image of yourself. Even if you have the ultimate knowledge in all architecture domains, do not compromise the appearance of your diagrams at any cost. Modeling is an art, and your diagrams will carry your signature for as long as the enterprise architecture repository is alive, so make sure that you are putting your name on the most beautiful artifacts that you can make.

> **Important Note**
> In a nutshell, never compromise on making outstanding diagrams at any time. To make outstanding diagrams, you must pay attention to the details of how they look.

Have look at *Figure 3.16* in *Chapter 3, Kick-Starting Your Enterprise Architecture Repository,* and pay attention to the following details in case you did not previously notice them:

- All business Actors are the exact same size, separated by the exact same spacing, and they are all aligned to the top edge.

- The **Shipping Manager** business Actor, the **Vehicle Tracking** application service, the **Tracking App** application component, and the **Device Location** technology service are all aligned to the same vertical centerline.

- The **Note** element and the **Tracking App** component are aligned to the same horizontal centerline.

- The **Tracking App Context Diagram** label and the **Note** element are aligned to the same left edge.

- The **Vehicle Tracking** service and the **Trading Web** component are aligned to the same bottom edge.

- The **Vehicle Tracking** service and the **Device Location** service are the exact same size. Even though they are of different types, it looks nicer to have them of the same size.

- The spacing between the business Actors, the **Vehicle Tracking** service, and the **Tracking App** component is exactly the same.

- All **Business** architecture, **Application** architecture, and **Technology** architecture elements use the same style. Only the focus elements have slight differences, and that is for a reason.

- Simple colors are easy on the eye. If you want to use different color styles or even black and white diagrams, keep them nice to look at, and remember that not everyone favors the same colors that you do. Notice that the examples provided used only four colors and only for element borders. This provides a subtle difference when viewing the diagrams in color or black and white.

- Relationship lines should not cross each other. Sometimes, crossing lines is unavoidable, but it is always best to avoid this whenever possible. You may need to reorganize elements by swapping their places to avoid—or at least reduce—crossed lines. Minimizing the number of elements on a diagram helps to alleviate crossed relationship lines. Minimizing the number of elements can often be accomplished by focusing the diagram on a single idea. In the case of the **Tracking App Context Diagram** example, the focus was the context of a single component—**Tracking App**. While there may be several other components related to **Trading Web**, we avoid placing them on this diagram because they don't support the main idea.

- Related elements (such as business Actors) are visually grouped together. If there were other application components on the diagram, for example, we would have put them next to the **Trading Web** component and we would have made them all the same size and aligned them. The same applies to the technology service and any other element that we will use.

Now, look back at *Figure 4.9* too and see whether the preceding guidelines have been followed. You need to keep this advice in your mind always, no matter whether your diagram has 2 elements or 20, whether your diagram will be shared with your coworkers or with the **chief executive officer** (**CEO**), or whether it is a draft version or a final version. Sparx provides a fair number of sizing, spacing, and alignment tools in the **Layout** > **Alignment** toolbar that will make your job easier, so use them effectively. We apologize if we are repeating ourselves, but please treat your diagrams as artworks always carrying your name.

Knowing your audience

This is the most important and the most obvious rule that you need to follow in every diagram, presentation, and publication you write. This rule is not mine, in fact, and every person who is responsible for building and developing products knows it, whether the products are books, electronic devices, vehicles, or houses. In each of those cases, you must know your target audience, or else you will end up building something that no one is interested in.

In your case as enterprise architects, if you are targeting a C-level audience, you need to provide completely different content than what you would do if you were targeting technology gurus or professionals for a specific product. It is extremely important to realize that your job is to close the gaps of understanding among the different stakeholders and to avoid creating new ones. Creating diagrams—or creating architecture content in general—that do not address the concerns of the target audience will create new gaps for sure, and the least damage you would cause will be your wasted time and the additional confusion that will arise as a result.

Let's recap what we have learned in this chapter before moving on to the next one, where we will create more detailed diagrams showing other aspects of the application.

Summary

Using best practices and maintaining consistency in your architectural artifacts is very important to gaining and maintaining trust in your enterprise architecture practice. Consistency breeds familiarity, and with familiarity comes trust. We have learned how to interpret standard metamodels and how to build our own focused metamodel. From this point forward, when we need to describe another application component in a diagram, we will use a focused metamodel to guide us in knowing which elements can be placed on the diagram and which types of relationships can be defined.

As we introduce you to new artifacts and component types in this book, we will accompany that new information with a new focused metamodel. More metamodels will help us to describe more elements of the enterprise in useful diagrams. With more enterprise elements properly described, we will build trust in the enterprise architecture practice that we are establishing within our organizations. This is how you can build a practical enterprise architecture practice that provides actual value to the enterprise.

In the next chapter, we will continue adding content to the enterprise architecture repository to show more aspects and different views of the **Tracking App** application, so keep moving before you lose momentum.

5
Advanced Application Architecture Modeling

To develop or procure a business application, it is not sufficient to stay at a conceptual level of detail. The more detail that you provide, the easier it becomes for developers to build or customize a solution according to your requirements. However, as an enterprise architect, you are not expected to provide details for the developers because that is the job of a solution architect. Your job, in fact, is to provide details for the solution architects themselves to have the best understanding of the business requirements and direct the application, data, and technology architects on what will be delivered and how.

In this chapter, we will describe the **Tracking App** application from various perspectives. The two types of perspectives you can choose to describe an application are the structure and the behavior, and this is what we will be learning about in this chapter. We will be covering the following topics:

- Determining what diagrams to produce
- Describing application behavior
- Describing application structure

There are more diagrams to explore in this chapter compared to previous chapters, so get ready for a faster pace and more condensed information. You are not new to Sparx anymore, so you should be just fine.

Technical requirements

This chapter does not have any additional technical requirements other than having **Sparx Systems Enterprise Architect** (**EA**). If you do not have a licensed copy, you can download a fully functional 30-day trial version from the Sparx Systems website (`https://sparxsystems.com/products/ea/trial/request.html`).

We will continue adding the content of this chapter in the same EA repository that we built in *Chapter 3, Kick-Starting Your Enterprise Architecture Repository*, and *Chapter 4, Maintaining Quality and Consistency in the Repository*. If you have not read these two chapters, we strongly advise you to read them first and then come back to read this chapter, as many step-by-step instructions that have been detailed in these two chapters have been skipped or summarized in this chapter to avoid repetitive information here and there.

If you want to practice while reading, you can download the repository file of *Chapter 4, Maintaining Quality and Consistency in the Repository*, from GitHub at `https://github.com/PacktPublishing/Practical-Model-Driven-Enterprise-Architecture/blob/main/Chapter04/EA%20Repository.eapx` instead of starting from scratch. Some of the steps in this chapter depend on elements that have already been created in the repository, so it is better to not start this chapter with an empty repository. However, if you want to download the resultant repository of this chapter, which contains all the diagrams that will be created in it, you can do so by following this link: `https://github.com/PacktPublishing/Practical-Model-Driven-Enterprise-Architecture/blob/main/Chapter05/EA%20Repository.eapx`.

We will use the following **ArchiMate® 3.1** specification chapters to guide our development:

- *Chapter 5, Relationships* (`https://pubs.opengroup.org/architecture/archimate3-doc/chap05.html#_Toc10045310`)

- *Chapter 9, Application Layer* (`https://pubs.opengroup.org/architecture/archimate3-doc/chap09.html#_Toc10045389`)

- *Chapter 12, Relationships Between Core Layers* (`https://pubs.opengroup.org/architecture/archimate3-doc/chap12.html#_Toc10045440`)

The metamodel diagrams in the aforementioned references will be used throughout this chapter, so we highly advise you to print them and keep them in reach for maximum benefit.

Determining what diagrams to produce

Before we get into the various types of diagrams that you can produce, we need to introduce a couple of concepts, the concepts of the *view* and the *viewpoint*. If you're involved in a project where a single diagram is supposed to explain everything you need to know about a system or application, then drop the diagram, turn around, and run. Such a diagram cannot exist.

Some diagrams claim to do so, but they are usually far too big, far too busy, and don't convey a quarter of what is needed. They usually follow no standard and are created by those with little or no modeling experience. They make little sense except to their creator. The biggest problem with such diagrams is that they tend to convey a single set of concerns, those of the creator.

But please be careful, and be kind! These folks may have expended a lot of time and effort. They are often very knowledgeable and very proud of their work. The last thing you want to do is to create enemies on the project. Instead, praise their work and bring them into the fold of architecture professionals. Hand them a copy of this book. Also, there is often some useful information in these diagrams – information that you can glean for more meaningful models.

Understanding the view

So, how do you know what diagrams to create? Any given diagram or set of diagrams can be considered a **view** of the system in question. Every system has many stakeholders with their own view of the system, so you may have to develop a different number of views for each stakeholder. Each view may convey a slightly different set of information. Hence, which views should be created depends on the stakeholders and their concerns.

You may need to create many views of the system to address the concerns of all the stakeholders. This implies that you need to know who the stakeholders are, and what their concerns are. Often, you can infer a great deal about their concerns based on their position in the organization. For example, the controller, accounting manager, and bookkeepers are more likely to be interested in the financial aspects of the system. They will want to know about financial reports created by the system as well as financial records created or manipulated by the system.

In contrast, computer operators want to know about things such as backup, restore cycles, exceptions, batch processing, or anything that may affect them. Developers want to know what components make up the system and each of their roles, what services are expected, and so on. Views can mainly be categorized into two main categories, **structural** and **behavioral**, as we will see next.

Structural and behavioral views

We will look at the types of diagrams you can create, but let's first talk about the difference between structure and behavior views. We will use the human body as an analogy to clarify the difference. The way that we react when we hear sounds, for example, describes behavior. We may react to music in one way and to conversations in a different way. We react to loud sounds of danger in yet another different way.

Our bodies receive sound waves from the environment through our ears. Our brains translate that information and react by behaving accordingly. In fact, the process of receiving sound and transmitting it to our brain is also part of the body's hearing *behavior*. To understand the hearing function, we need to understand the parts of the body that play a role in hearing and how they are related. The ear, ear drum, ear canal, cochlea, cochlear nerve, and the brain all play a role in hearing. How they are related to each other and their individual roles comprise the *structure* of our hearing ability.

If we were to describe hearing ability in a set of models, we would likely describe the overall structure of the body at a high level to identify where the ear is located, and then, in a separate diagram, we would describe the components of the ear. These two views would describe the structure of our hearing ability. To describe the process of hearing, we would need yet another set of diagrams.

An overall diagram would describe the overall path of a soundwave as it enters the ear canal and gets translated into an electric signal that enters the brain. A lower-level set of diagrams might be needed to describe the process each component of the ear uses to accomplish its individual role. To completely understand how we hear, we need both the structural and the behavioral diagrams. Most importantly, we need to know what our target audience is interested in seeing in our diagrams, and this is where viewpoints play an important role.

Getting to know the viewpoint

Describing the hearing system for a sixth-grade student is different than describing it for a family doctor who specializes in the **Ear, Nose, and Throat** (**ENT**). The details that you put in both books will be completely different even though they both talk about the same system. They are both right, but each addresses a different audience and serves a different purpose. This is where viewpoints come in – to tell us what each stakeholder is interested in.

If you're in an organization and working with a team of architects, it may be helpful for you to create what is known as a **viewpoint**. This is simply a specification for the view. It describes what type of diagrams you will create, what notation you will use, what stakeholders you are targeting, and what concerns you are addressing. Basically, you are telling the reader what they should be getting from the view. The concepts of the view and viewpoint are specified in **TOGAF® 9.2** (`https://pubs.opengroup.org/architecture/togaf9-doc/arch/chap31.html`).

There are also several predefined viewpoints available with the standard. The specifics of how to create a viewpoint are beyond the scope of this book. We encourage you to look at the standard and determine for yourself the degree to which you need to specify a viewpoint for your views. Suffice to say that considering the readers (stakeholders) of your diagrams and their concerns is probably the most important thing you need to ponder.

Once you know what your stakeholders are interested in seeing, it is time to produce some diagrams describing the Tracking App application component, and we will start by describing the behavior of the application.

Describing application behavior

Application behavior describes the functions, processes, and services an application delivers. Behavior includes answers to questions such as what happens, in what order, what is used, and what the result is. Behavior also describes what functionality is made available to other components as services, and what happens when an unexpected event occurs. This information can be modeled and provided as a way of documenting the application, so application stakeholders can visualize the application behavior and make early decisions regarding it. This is what we will learn about in this section as we answer the following questions:

- What are application services, who needs to know about them, and how do we model them?

- What are application processes, who needs to know about them, how are they related to the application services, and how do we model those processes?

- What are application functions, how are they related to services and processes, and how do we model them?

- What are application events, and how do we model them?

Keep in mind that the elements mentioned in the preceding list – services, processes, functions, and events – all represent behavior. In some cases, they can overlap, which confuses many architects and causes continuous disagreements.

What one architect perceives as a function might be perceived as a process by another architect and as a service by a third. They could all be right, but because of their different experiences and the different stakeholders they serve, they do not necessarily agree. We will explain the differences with examples and learn how to build the right models that convey the right level of detail. Let's start with application services.

Introducing application services

According to ArchiMate® 3.1, an **application service** *"represents an explicitly defined exposed application behavior"* (https://pubs.opengroup.org/architecture/ archimate3-doc/chap09.html#_Toc10045401).

This means that application services are clearly visible to the external enterprise, so they are what the enterprise sees and knows about a specific application. When looking at an application service catalog, you are looking from the external *service consumer* perspective.

ArchiMate® provides two notations to model application services – rectangular and borderless. We will use both notations in some different diagrams, but in the end, it is a matter of preference:

Figure 5.1 – ArchiMate®'s Application Service notations

Knowing what services an application component provides is different than knowing how to access and consume the services. Thus, it is very important to understand that application services describe the behavior of an application component, not its structure. The part of the application that you use to consume a service, such as exchanging data or performing an action, is called the **application interface**. We will talk about it in more detail in *Introducing application interfaces* in the next section of this chapter. **Interfaces** are the structures that provide access to services, so they are the *exposed tangible* part of a component. They are the entry points that allow consumers to use services.

Another very important thing that you need to understand is the difference between application services and application components. The line between the two is very thin, but it is very clear once you see it. Architects coming from a **Service-Oriented Architecture (SOA)** background might call the entire component a service or **web service**. Web services, from ArchiMate®'s point of view, are application components, not application services, because they encapsulate functionality, and are modular and replaceable.

For example, if you have a web service that prints mailing labels, then ArchiMate® considers it an application component that provides the printing mailing labels service. So, there are two types of architectural elements that we can use to model our web service example. The application component is the *tangible* and deployable part, while the application service is the *descriptive* part. Please remember this essential difference and remember that it is a quite common confusion because of how other industry standards define services. We are following ArchiMate®'s definitions in this book.

A single component can provide multiple services, so it depends on how the component is designed and constructed. Some components can provide hundreds of services, while some microservice components may provide only a single service. The number of provided services does not affect the classification of what a service and a component are. A single service can be provided by multiple components as well. For example, you may have multiple products from multiple vendors, each of which prints mailing labels. They may differ in how the service is performed and provided, but they all provide the same service. Also, multiple components may work together to provide a single service.

Additionally, we need to differentiate between application services, technology services, and business services. They obviously belong to different architecture layers, as their names imply, but as you categorize the services in your enterprise, it may become difficult to decide which service belongs to which layer. **Business services** are the services that your business provides to its customers regardless of whether they are done through a website, a call center, or a sales showroom. A grocery store, for example, provides a retail grocery sales business service whether you walk in, use their online virtual store, or order by phone. We will talk in more detail about business services in *Chapter 8, Business Architecture Models*.

Technology services are usually the services that are provided to your business at the technology infrastructure level, such as networking, storage, protection, computing, and many other services. Services that are available on **Amazon Web Services (AWS)** or **Microsoft Azure** are great examples of technology services, as they are not related to any specific business type. We will talk in more detail about technology services in *Chapter 6, Modeling in the Technology Layer*.

Finally, application services are software services just like technology services, but they encapsulate business-specific logic in them, so they are not as business-neutral as technology services. They also encapsulate technology, so they are not as technology-neutral as business services. They are in between the two. They *apply* technology to help solve business problems, thus the name *application*. Order processing is an example of an application service that automates processing orders for a specific retail business, and to do that, it uses business logic and rules that are specific to that business. All business types process their orders differently, so each can implement the order processing application service differently.

We need to look at some real-world examples to clarify all these points about application services, but let's first define the focused metamodel that we will use to guide the development of our models.

Defining the application service focused metamodel

Focused metamodels are models that guide the development of a single specific element at once. Each focused metamodel will have a single focus element that describes every possible connection between it and other enterprise elements to which it is allowed to connect. The focused metamodels are simply an element-by-element translation of the ArchiMate® 3.1 metamodels. If you were to combine all these focused metamodels, you would get the complete metamodel of all ArchiMate® 3.1 elements. It would be a very complex diagram, with many intersecting and overlapping connectors. It would be difficult to read, so we will not build it. Just remember that it can be built.

In *Chapter 4, Maintaining Quality and Consistency in the Repository*, we developed a focused metamodel for the application component element, and we saw how the application component was at the center (the focus), while other elements surrounded it. We will use that focused metamodel whenever we need to build a model describing an application component.

In this section, we will develop the focused metamodel for the application service element, and we will use it whenever we build a model describing an application service. The application service element will be the focus this time, while other enterprise elements including the application component will surround the application service element. The application component that was the focus in *Figure 4.9* of *Chapter 4, Maintaining Quality and Consistency in the Repository*, will appear in the application service focused metamodel as a secondary element, and the application service element that appeared as a secondary element in that diagram will appear as the focused element in this focused metamodel. However, it is very important to keep in mind that they are the same elements, regardless of which one is the focus and which one is not.

Think about it as if you are taking pictures of your children on their graduation day. Most likely, you will focus on them in all the photos you take, while every other student in the background will be secondary person. Your children will also appear in other parents' pictures, but they will be a secondary persons in their pictures because their child will be the person focused on. In both cases, your child is still the same person, and if the school wanted to count how many students appeared in all the pictures taken by all parents, every student would be counted once, regardless of how many pictures they appeared in.

We will use the ArchiMate® 3.1 standard to guide our focused metamodel development, and we will translate our understanding and interpretation of it into an easier-to-understand reference for every architect that will contribute to the enterprise architecture repository.

Interpreting the standard

The following list is our interpretation of the diagrams and definitions in the three reference chapters that we indicated in the *Technical requirements* section:

- An application service *serves* an application's internal behavior element (such as an application function, an application process, or an application interaction).

- An application service can be *realized by* an application's internal behavior element.

- An application service *serves* an application's internal active structure element (such as an application component or an application collaboration).

- By default, every enterprise element can be composed/decomposed into elements of its type, so an application service can *compose/decompose* other application services.

- An application service can *trigger*, *be triggered*, or *flow* to another application service.

- An application service can *trigger*, *be triggered*, or *flow* to an application event.

- An application interface can *be assigned to* an application service.

- An application service can *access* a data object.

- An application service *realizes* a business service.

- An application service *serves* a business internal active structure element (a business role, a business actor, or a business collaboration).

- An application service *serves* a business internal behavior element (business process, business function, and business interaction).

- An application service is *realized by* a technology service.

- An application service *serves* a technology internal active structure element (a node or a technology collaboration).

- An application service *serves* a technology internal behavior element (a technology process, a technology function, or a technology interaction).

As you can see, the aforementioned list of statements about the application service metamodel is a bit long, which is expected for an element that is externally exposed to other enterprise elements. As a result, we expect the resulting diagram to be large and contain many elements, so just keep that in mind. We have sufficient information to start building the application service-focused metamodel, so let's get ready.

Building the focused metamodel

The application service will obviously be the focus element in the application service-focused metamodel. But before you drag a new application service element from the toolbox to the diagram, remember that our repository is no longer brand new, and we might have already created that element.

If you revisit *Figure 4.9* in the previous chapter, you will see that we have already created an application service element in that diagram, which means that it already exists in the repository, and we must reuse it in this and other metamodel diagrams. Follow these steps to learn how:

1. From the project browser, open the **Metamodels** package, create a new diagram in it, and name it `Application Service Focused Metamodel`.

2. Optionally, choose to show diagram details (the blue label at the top-left corner of the diagram). Also optionally, change the diagram's theme to **High Contrast White** to remain consistent with what we are building.

3. Find the application service element in the metamodel package and place it on the diagram. Sparx proceeds every ArchiMate element in the repository with a stereotype that follows this pattern – *<<ArchiMate_element type>>element Name*. So, the full name of the element that you are looking for is **<<ArchiMate_ ApplicationService>>Application Service**.

4. When you place an existing element on a diagram, Sparx may display a dialog box asking whether you want to reuse the item as a link or a child from the **Drop as** drop-down list:

Figure 5.2 – The Paste Application Component dialog box

This dialog box will not be displayed by default for every element type, so to force it to appear, hold the *Ctrl* key while dragging.

5. Select **Link** because this tells Sparx that we are reusing the same element on a new diagram. Choosing **Child** means that the copied element is a specialized child of the existing element, a technique that is particularly useful when modeling classes in the Unified Modeling Language.

6. Leave the remaining options as they are and click **OK**.

    The next step is to add the related elements as interpreted by the ArchiMate® specification. Remember that some of these elements already exist in the **Metamodels** package, so you must check to see whether the elements to be added to the diagram already exist or need to be created.

7. Find the following elements in the **Metamodels** package and place them on the diagram: application event, application process, application function, application interaction, application component, application collaboration, application interface, data object, business service, and technology service.

    Once you place the application component element on the diagram, Sparx displays all the relationships that already exist between the application component and the other elements. The same thing will happen every time you place an existing element on the diagram, and all the relationships between that element and the other visible elements on the diagram will be shown too.

    Seeing these relationships in other focused metamodels and diagrams is useful, as we can understand all the possible relationships to and from the elements. However, our focus element, for now, is the application service, so to keep it focused, the application component relationships are irrelevant to its context. Remember that the main objective of creating the focused metamodels is to focus on a single element at a time, so we need to hide the needless relationships.

8. To hide a relationship from a diagram, right-click on the desired relationship and select **Visibility** > **Hide Connector**. Alternatively, to hide a relationship, highlight it by clicking on it, press the **Delete** button, and a dialog box like the one shown in the following screenshot will pop up, asking you whether you want to permanently delete the relationship connector or just hide it in this diagram. Since we only want to hide the relationship from this diagram, keep the **Hide the connector** choice selection, and click the **OK** button:

Figure 5.3 – The Remove Connector popup

9. Repeat the previous steps for all the relationships that do not connect to or from the application service element because, in this focused metamodel diagram, we only need to show the application service relationships.

> **Important Note**
> Another useful way to show or hide a long list of relationships in a diagram is to click **Layout** > **Diagram** > **Appearance** > **Set Visible Relationships**. Sparx will show you a pop-up window containing a list of all the relationships that exist between the elements on the diagram. Check the box next to the relationships that you want to show and uncheck them for the ones that you want to hide.

There are still elements that have been identified in the interpretation list, but they do not exist in the **Metamodels** package because we have not created them in the repository yet. For these elements, they must be created from **Toolbox**, which will require activating the business toolbox and the technology toolbox to get the desired business and technology elements respectively. Refer to the *Adding Actors* subsection, in the *Adding elements to the diagram* section, from *Chapter 3, Kick Starting Your Enterprise Architecture Repository*, to see how to change the toolbox, if you forgot how.

10. From the business toolbox, add the following elements to the diagram: business role, business actor, business collaboration, business process, business function, and business interaction.

11. From the technology toolbox, add the following elements to the diagram: node, technology collaboration, technology process, technology function, and technology interaction.

12. Optionally, style all application architecture elements using the saved **App Arch** style, all the business architecture elements using the saved **Biz Arch** style, and all the technology architecture elements using the **Tech Arch** style.

13. Optionally too, change the appearance of all the elements to the borderless notation while keeping the application service element with the rectangular notation.

14. Rearrange the elements on the diagram to have the application service in the middle, bigger than the other elements, with a thicker border, and surrounded by all the other elements.

    The last thing that we need to do to complete this focused metamodel is to create the relationships that connect the application service element with other elements, as we have defined in the interpretation. We already have the *serves* relationship between the application service and the application component, so follow the next steps to add the remaining relationships.

15. Create a *triggers* relationship from the application service element to itself to indicate that an application service can be triggered or triggered by another application service.

16. Create a *serves* relationship from the application service to the application process element.

    Remember that these two elements exist on another diagram, which is the application component-focused metamodel. If you open that diagram now, you will see that this newly created relationship is also visible there, which is not what we want. If we add all the relationships that we have on our interpreted list, we will turn the application component-focused metamodel diagram into a *spaghetti* diagram, with dozens of relationships going in every direction. Therefore, we must ensure that any new relationship that we create on this diagram will not be visible on other diagrams.

17. Right-click on the relationship connector and choose **Visibility** > **Hide Connector in Other Diagrams**. The **Set Connector Visibility** pop-up window will appear with a list of all the other diagrams that this relationship will appear on:

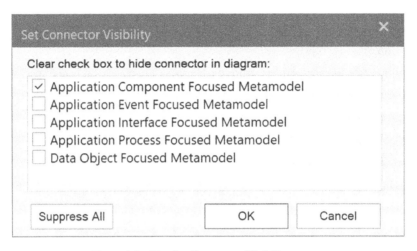

Figure 5.4 – The Set Connector Visibility popup

Important Note

Your list may contain more or fewer diagrams in it, based on the number of diagrams that contain the selected relationship.

18. Uncheck all the checkboxes that you want to hide the relationship from, and then click **OK**.

    Remember to keep the relationship visible on the diagrams that it needs to be visible on. If you hide all the relationships in all the other diagrams, you will end up with only the active diagram showing relationships, which is not what we need to achieve.

19. Repeat *step 17* to *step 18* for all the relationships that are identified in the interpretation list.

    In some cases, even if we were careful in showing and hiding the relationships in other diagrams, we may accidentally hide a needed relationship and show an unneeded one, so we recommend you go back to older diagrams and check to make sure that they still look okay.

20. Open the application component-focused metamodel diagram and make sure that all the relationships that we want to keep visible are visible, and all the relationships that we need to hide are hidden.

21. Finally, make sure to save your work.

This was a difficult metamodel to create, so don't feel that you're the only one who faced difficulties in understanding it or struggled in following the steps to create it. In *Chapter 4, Maintaining Quality and Consistency in the Repository*, we created a relatively simply focused metamodel and provided more detailed instructions than the ones we provided here, so refer to it for more information.

> **Important Note**
>
> Creating metamodels in your repository is important but not essential. You can use the metamodels that are provided in this book, or you can use the standard ArchiMate® metamodels to create diagrams. Enriching your repository with diagrams is more important than enriching it with references.

By following the aforementioned steps, you will get a diagram like the following:

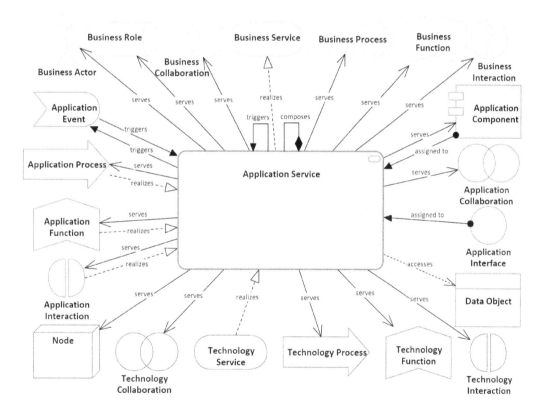

Figure 5.5 – The Application Service-focused metamodel

As expected, the **Application Service**-focused metamodel contains many elements and relationships, but congratulations – this will be a reference in your repository that you will rarely change! Barring a few cases where a new version of the ArchiMate® specification is released, you find a missing element or a mistake, or the organization that you work for has a specific need to adjust or override the standard, this diagram is ready to be printed and pinned to your workspace. You can refer to it whenever you model application services.

Use this focused metamodel to build an unlimited number of enterprise architecture models, and it will be your guide to maintaining consistency in your repository as well as adhering to the industry standard.

In the next subsection, we will show you some examples of how we can build diagrams based on the focused metamodel that we have just defined.

Modeling application services

In *Chapter 3*, *Kick-Starting Your Enterprise Architecture Repository*, we developed a model that identifies what a Tracking App is, what services it provides, and other similar information. In this subsection, we will elaborate more on the provided **Vehicle Tracking** service, so we will develop more models that have it as the focus element. Remember that these are just examples to convey the idea of different model types. You need to apply these examples in your work environment with actual components from your enterprise and based on the concerns of your stakeholders.

Modeling application service context

Context diagrams show an element in the *context* of its environment or surroundings. These diagrams apply to any element in the enterprise, as they show multiple aspects and relationships all in one place. They provide an overview, not a detailed view. You can always provide more detailed models when needed. We introduced the Tracking App to the stakeholders in *Figure 3.16* of *Chapter 3*, *Kick-Starting Your Enterprise Architecture Repository*, using a context diagram. We will drill down from that diagram and provide a more detailed model just about the Vehicle Tracking application service. The purpose of this diagram is to give a more detailed overview of how this specific application service fits into its environment.

We will use the application service-focused metamodel from *Figure 5.5* as a reference for developing this model and knowing what elements can relate to the application service. There are 21 different element types on this metamodel, but this does not mean that we must use all of them in our context diagram. We will use only what we need to answer the question, *what is the vehicle tracking service?* If you want, you can print a copy of the application service-focused metamodel, and follow these steps to create a context diagram for the Vehicle Tracking application service:

1. Open the **Tracking App** package in the project browser.

2. Right-click on the package and select **Add Diagram** to open the **New Diagram** dialog box.

3. Enter `Vehicle Tracking Context Diagram` as a name for the diagram. Select the **ArchiMate3.1 Application** diagram type.

   Since we already created the Vehicle Tracking application service earlier, we need to reuse it in this diagram and expand it by adding more information to its context.

4. Locate the Vehicle Tracking application service in the **Tracking App** package and place it in the center of the new diagram.

   Since this is an existing element and we are reusing it, it may also contain some of the information that we need to have in this new context diagram. Therefore, we need to check what information is available and decide what we need to reuse.

5. Right-click on the **Vehicle Tracking** element on the diagram and select **Insert Related Elements** from the context menu. This will open the dialog box that is shown next:

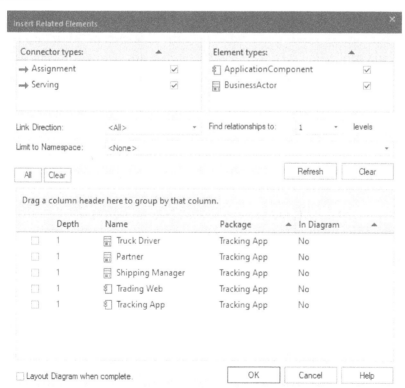

Figure 5.6 – The Insert Related Elements dialog box

The dialog box shows all the items that are related to the selected element, which in our case is the Vehicle Tracking service. You can select or unselect the elements that you want to insert. You can also filter by **Connector types** or **Element types** and then refresh the list.

6.   In this diagram, we want to reuse all the related elements, so click **All** to select all the listed elements, and then click **OK**.

7.   The inserted elements are put on the diagram as a cascaded list, so reorganize them in a way that makes sense to you, and style them if needed. Also note that the related elements are inserted with the relationships that we created earlier in *Chapter 3, Kick-Starting Your Enterprise Architecture Repository*.

One important aspect of application services, in general, is that they exist to realize business services. If an application service does not realize a business service, then it is considered an extra or even a useless service that an application has but the business does not use. The beauty of enterprise architecture is that it shows you all these essential connections. In this context diagram, we need to show the business service or services that are realized by the Vehicle Tracking application service.

Business services will be covered in detail in *Chapter 8, Business Architecture Models*. For now, we need you to understand the difference between business services and application services. Business services are *neutral* from any application or technology implementation. They exist with or without computers. Automation and technologies make them more efficient for sure, but the existence of business services is an essential part of performing business strategic missions themselves. For a trading company, tracking vehicles' locations is not a service that they need to provide, but goods transportation is. Therefore, we can say that vehicle tracking is an application service that realizes the goods transportation business service and makes it more efficient.

Furthermore, the *ABC Trading* company may have an **Enterprise Resource Planning** (ERP) solution that provides a service for goods transportation. This will be the **Goods Transportation** application service realizing the **Goods Transportation** business service. Even though the two services can have the same name and are tightly related to each other, it is very important to differentiate them as two architectural elements when they are defined in the repository.

Having said that, we need to add this realization to the model in the next step.

8.   Change **Toolbox** to **ArchiMate 3.1** > **Business**, pick the **Business Service** element, drop it on the diagram, and rename it `Goods Transportation`. Style the business service element if you want.

9. Create a *realizes* relationship from the application service to the business service.

    For an application service to work, it needs data and a set of processes to manipulate it. Let's add these two elements to the diagram. We know that this service will provide the device location data to consumers, so we need to add it to our diagram.

10. Find the **Device Location Data** object in the **Tracking App** package and drop it on the diagram. Style it if needed.

11. Create an *accesses* relationship from the application service to the data object.

    Now, we need to define the process or the processes for manipulating the device location data. We will talk in detail about application processes later in this section, so let's keep things simple for now and add a new application process to the diagram.

12. Change to the **ArchiMate 3.1 Application** toolbox if it is not already active.

13. Take a new application process element, place it on the diagram, and rename it to `Manipulate Device Location Data`.

14. Create a *realizes* relationship from the application process to the application service.

    Finally, we need to indicate how this application service will be provided. In other words, what application interfaces will provide the service? Services are provided to users through **User Interfaces** (**UIs**) and to applications through **Application Programming Interfaces** (**APIs**). An application service can be provided by many application interfaces, and an application interface can provide many application services. We will talk in detail about application interfaces in the next section of this chapter, but for now, follow the remaining steps.

15. Find the application interface element in the toolbox and use it to add six new application interfaces to the diagram.

16. Name the application interfaces `Android UI`, `iOS UI`, `Command Line UI`, `Web UI`, `SOAP API`, and `REST API`.

17. Create *assigned to* relationships, going from the application interfaces to the application service.

> **Important Note**
> Always remember that our references for developing models (or diagrams) are metamodels. Our reference metamodel indicates that the relationship between an application service and a data object, for example, is of the `accesses` type, so we must adhere to this and avoid creating any different relationships that are not part of the standards.

The diagram has enough information by now, but this does not mean that we cannot add any additional elements to it. We can always come back and add more elements as we discover them. We will keep learning, and our stakeholders' requirements will keep evolving too. The diagrams for your actual work may end up bigger, more complicated, and with more elements. If they become too busy, then it is an indicator that you need to split the information over multiple smaller diagrams. Refer to the modeling guidelines that we provided in *Chapter 4, Maintaining Quality and Consistency in the Repository*.

For the sake of the example, we will stop at this amount of information. The **Vehicle Tracking** context diagram should now look like the following:

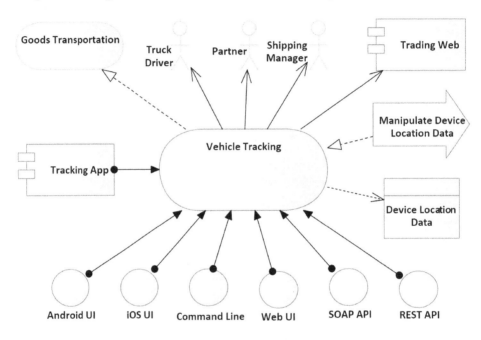

Figure 5.7 – The Vehicle Tracking application service context diagram

This artifact may look very different from what you were expecting for an application service model. But this way, you are presenting the application service at a conceptual level that makes it perfect for management to understand. Always keep the targeted reader in mind.

Even though it represents an application-level artifact, it is void of implementation detail, which is necessary for management-level stakeholders. You can add a deeper level of detail in a different diagram if you choose to and have the experience. You can also hand this conceptual design to another architect who can provide that detail. Remember that your role as an enterprise architect does not mean that you must build everything but instead make sure that the enterprise elements and stakeholders are connected and communicating properly. A diagram such as the one shown in *Figure 5.7* will help to ease these connections.

The next subsection will provide you with another example of a management-level artifact that you can add to your enterprise architecture repository.

Modeling an application services catalog

Context diagrams are useful, but they show one element at a time. What if you want to show multiple application services in one diagram? The **application services catalog** is an artifact that shows a list of application service elements. The size of the list varies based on its scope.

You can have a catalog of all the application services that are provided by a single application component. Another catalog might contain a list of all the application services in the enterprise. A third catalog might contain a list of all the application services that a given business actor is looking for in a targeted (or to-be targeted) application, and so on.

The following diagram is an example of an enterprise application services catalog. Depending on its scope, your actual catalog will most likely be larger:

Figure 5.8 – An enterprise application services catalog example

Note that in *Figure 5.8*, we have only used one of the relationships that were identified in the application service-focused metamodel, which is self-composition. Services can be composed of smaller services, which in turn can be composed of yet smaller services. If you are modeling a set of services that already exist in the enterprise, it is a good practice to keep them all together in the same package structure.

This makes it easier when you need to reference the service in future models. The services package may contain sub-packages to keep the services organized. If, on the other hand, you are working on a specific project for a new application that has not yet been developed or implemented, you'll probably want to keep the services in a project-specific package. Either way, it pays to keep your repository organized. It's also a good practice to revisit the subject of repository structure from time to time. An occasional restructure is not uncommon.

Linking diagrams in Sparx

In some cases, you may have a high-level diagram that drills down to one or more detailed diagrams. We have *Figure 5.8*, which lists all application services provided in the enterprise. We also have *Figure 5.7*, which details one of these application services. Being able to link the two diagrams will provide users of the enterprise architecture repository the ability to navigate from one level of detail to another, or from one element to another. This is a very powerful feature in Sparx, and by using it properly, your enterprise models can all be connected. Users will also have a better understanding of the enterprise components by navigating through models.

To link *Figure 5.7* and *Figure 5.8*, follow these steps:

1. Open the **Enterprise Application Services Catalog** diagram that we reviewed earlier in this section.

2. Locate the **Vehicle Tracking** application service on the diagram. Right-click on it and select **New Child Diagram** > **Select Composite Diagram** from the context menu.

    The **Select Classifier** dialog box will open, asking you to select the child diagram of the selected element, as follows:

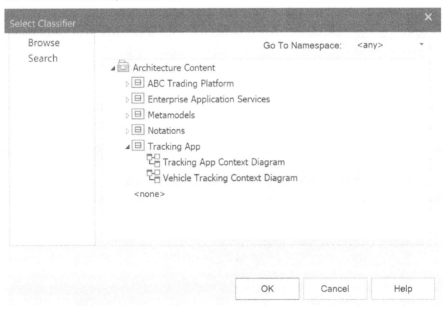

Figure 5.9 – The Select Classifier dialog

3. Find the desired diagram, which is **Vehicle Tracking Context Diagram** in our case, and click **OK**.

If your enterprise repository becomes too large and composed of many levels of nested packages, you may find it difficult to find the desired diagram by yourself. It will be better to search for the diagram and let Sparx find it for you, as explained in the following steps.

4. In the left tab of the **Select Classifier** dialog, click on **Search**, which will change the content of the dialog box to the following:

Figure 5.10 – Search for the classifier

5. Type the name of the diagram that you are searching for or type a small part of it, and then click **Find**. This will list all the diagrams that match the search criteria.

6. Select the desired diagram from the results list and click **OK**.

Note how the **Vehicle Tracking** application service has a small chain symbol at the bottom-right corner, as shown in *Figure 5.11*. This indicates that the element has a linked child diagram:

Figure 5.11 – The chain symbol indicates a linked diagram

> **Important Note**
>
> Elements modeled using the borderless notation will not show the chain symbol at the bottom-right corner of the element. Only rectangular elements will.

Double-click on the **Vehicle Tracking** element, and you will see how Sparx opens the linked child diagram in the diagram area. You can navigate back to the previous diagram by clicking the back arrow at the corner of the diagram tab, as shown in *Figure 5.12*. The back arrow replaces the standard **x** symbol that we usually use to close a diagram:

Figure 5.12 – The back arrow replacing the x symbol

Another way to know whether an element has a child diagram or not is to single-click on the element to highlight it. If the element has a linked child diagram, an eye-like symbol will appear above the top-left corner of the element, as shown here:

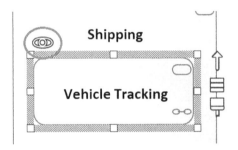

Figure 5.13 – The eye symbol indicates a linked child too

Clicking this eye will provide a read-only view of the child diagram in a pop-up window without opening it in the diagram area. As you can see, navigation between models is a very powerful feature. It can take you from one element to another and from one level of detail to another, just as hyperlinks work in web browsers. This is a very convenient way to help your enterprise architecture users understand more about the enterprise and how it is connected. Now, follow the same steps to link the Tracking App element to the Tracking App context diagram.

> **Important Note**
>
> You can link only one diagram to one element. However, you can make the same diagram a child of many other elements.

In the next subsection, we will learn about another application behavior element, which is the application process.

Introducing application processes

ArchiMate® 3.1 states that an **application process** *"represents a sequence of application behaviors that achieves a specific result"* (https://pubs.opengroup.org/ architecture/archimate3-doc/chap09.html#_Toc10045399).

Application processes are internal behavior elements. They describe how to achieve a specific result by executing a set of activities in a specific sequence. Suppose, for example, that the application needs a person's age in years. It can calculate the age by following these simple steps:

1. Get a person's birth date.
2. Get today's date.
3. Subtract the birth date from today's date and ignore the day and month in the result.

Application processes are always confused with business processes and technology processes, so we will try to highlight the main differences to avoid this confusion. They are all processes, to begin with, so they all perform a sequence of activities or steps to achieve a specific result. Business processes are neutral from any specific application or technology implementation. They exist with or without computers.

Every business, small or big, knows how to achieve specific results by executing a specific set of activities. For example, farmers knew how to plant and harvest for thousands of years before computers were invented. Traders used to buy and sell products and ship them overseas for hundreds of years. They knew how to plow the soil and plant seeds. They knew how to pack shipments, load them in proper containers, and deliver them safely to their destinations. All of these were business processes. Documented or not, they are still business processes. Automated or not, they are still business processes. Automation increases the efficiency of business processes, but the processes exist with or without it.

Automating a process means that we are building all or part of a business process inside a computer application. If we are building an application to automate the watering process, for example, we program the steps (or activities) of that business process in the application, and it results in what is categorized as an application process. Application processes may have the same names as business processes. We may have a business process named *Water the Corn Field*, and when we automate it, we may have an application process that has the same name, *Water the Corn Field*. They are still two different elements, even though they have the same name, the same activities, and they achieve the same outcome. They belong to two different architectural layers, and they represent two architectural elements. The application process element, however, realizes the business process element.

We can have multiple application processes realizing many business processes, so it does not have to be a one-to-one relationship. Additionally, an application process does not have to exactly match all the business process's sub-activities to realize it. You may still have some manual steps that a person needs to perform outside an application, such as signing a paper, checking the moistness of soil with an external device, or checking the weather forecast website.

In a fully automated business world, the business processes and the application processes may have a 100% match. In many realities, this match does not happen, but we can still consider that the application process realizes the business process. Business processes will be revisited in more detail in *Chapter 8*, *Business Architecture Models*, but you only need to know the difference for now.

Technology processes, on the other hand, are also processes, so they have a sequence, but they are at the technology level. This means that they are business-neutral, just like anything else at the technology level. Examples of technology processes include onboarding users, authenticating users and smart devices, encrypting content, performing a backup, archiving, publishing, compressing files, securing a network, and messaging. Technology processes include every process that IT departments perform to keep the lights on. The **IT Infrastructure Library** (**ITIL**) is a rich source of information technology processes and services, so we advise you to look at it if you are planning to define technology processes in your enterprise. Technology processes will be covered in more detail in *Chapter 7*, *Enterprise-Level Technology Architecture Models*.

To conclude this definition section, you need to remember that all types of processes have similar characteristics, but they differ in the layer to which they belong. Business processes are neutral from technology implementations, technology processes are neutral from business rules, and application processes are in the middle, connecting the two. Technology processes realize application processes, which realize business processes.

ArchiMate® provides two notations for modeling application processes, as depicted in the following diagram:

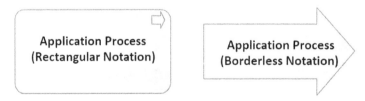

Figure 5.14 – Application Process notations

Now that you are familiar with the definition, let's build the application process-focused metamodel to guide us in developing our models.

Defining the application process-focused metamodel

Application processes are application internal behavior elements as per ArchiMate® 3.1, so the focused metamodel that we will build here is also applicable to other application internal behavior elements, such as application functions and application interactions. Therefore, we will only build this focused metamodel, and you can build the other ones as your own exercises.

The following list is our interpretation of the ArchiMate® 3.1 specification regarding application processes:

- Application processes can, by default, *compose* or *aggregate* other application internal behavior elements, such as application processes, application functions, and application interactions.

- Application processes *realize* application services, and application services *serve* application processes.

- Application processes *access* data objects.

- Application processes *trigger* application processes and application events, and application processes can be *triggered by* application processes and application events.

- Application internal active structure elements such as application components and application collaborations can be *assigned to* application processes.

- Application processes *realize* business processes.

- Application processes are *served by* business services.

- Application processes are *served by* technology services.

To build the application process-focused metamodel, please follow these steps:

1. Create a new ArchiMate® 3.1 application diagram in the **Metamodels** package and name it `Application Process Focused Metamodel`.

2. Reuse the application processes element from the **Metamodels** package, place it on the newly created diagram near the center, and make it the focus element.

3. Also reuse the following elements from the **Metamodels** package: **Application Function**, **Application Interaction**, **Application Service**, **Data Object**, **Application Event**, **Application Component**, **Application Collaboration**, **Business Process**, **Business Service**, and **Technology Service**.

4. Hide all the connections on the diagram except those connecting to the **Application Process** element.

5. Create the proper relationships between the **Application Process** element and the other elements on the diagrams if not created already.

6. Make sure to hide the newly created relationships on the other focused metamodels if needed.

Your final diagram should look like the following:

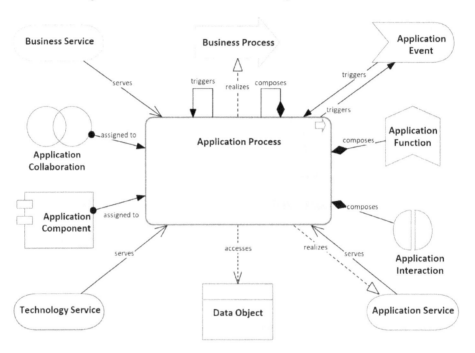

Figure 5.15 – The Application Process-focused metamodel

We hope that this focused metamodel will be a valuable reference for developing useful models for every stakeholder in your organization. We also hope that the example that will be provided in the next subsection will help you understand what level of application process models you can develop using ArchiMate®.

Modeling application process context

Context models in general are useful to inform viewers what a specific element is, how it works, and what it needs to work. In the Tracking App example, we know from *Figure 3.16* of *Chapter 3*, *Kick-Starting Your Enterprise Architecture Repository*, that the application gets location data from the mobile phone service and converts it into another form of data that will be useful to the enterprise consuming the service.

The device location data that comes from the mobile phone service might contain data such as the **International Mobile Equipment Identity** (**IMEI**), which is useful to uniquely identify a phone on a network, but this kind of information is useless to a person. What would be more useful for the person is to know the name of the driver that is carrying the phone. Therefore, the Tracking App must have the ability to manipulate raw device data and convert it into driver location data. In other words, the Tracking App needs to convert device location into data that can be understood by the business, such as driver location. The following example shows the **Manipulate Device Location Data** application process model:

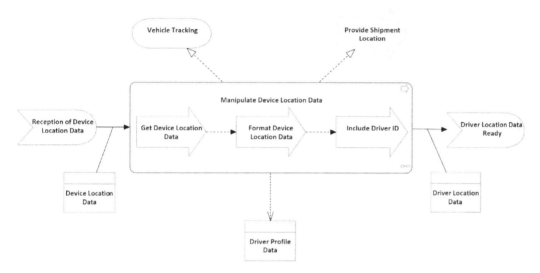

Figure 5.16 – The Manipulate Device Location Data application process diagram

As you can see, this diagram tells us what the process is, how it works, and what it needs to work. It tells us that it gets triggered by the reception of device location data, a sequence of child application processes performs a set of activities on the data to convert it into business meaningful data, and then an application event is triggered to inform the event listener that the data is available for consumption. The **Manipulate Device Location Data** application process receives the raw device location data, accesses the driver profile data, and produces the driver location data. Finally, the diagram tells us that the application process realizes the **Provide Shipment Location** business processes and realizes the **Vehicle Tracking** application service.

Now, we need to make this diagram a child diagram of the **Manipulate Device Location Data** application process element. If you've forgotten how, here is a quick reminder:

1. Right-click on the **Manipulate Device Location Data** application process element.
2. Select **New Child Diagram** > **Select Composite Diagram** from the context menu.
3. Select the diagram from the tree or search for it, and then click **OK**.

The repository users are now able to navigate to this context diagram from any other diagram that contains the **Manipulate Device Location Data** application process element.

Many people expect to see activities and decisions in a process model because this is how the **Unified Modeling Language** (**UML**) activity diagrams and **Business Process Modeling Notation** (**BPMN**) diagrams behave. ArchiMate® is an architecture language, and as such, it is not designed to replace UML or BPMN when it comes to modeling processes in detail. Most business and IT stakeholders are quite familiar with these standards, and ArchiMate® does not need to show complex flows with activities, decisions, forks, joins, and swim lanes. Most developers understand UML, and the application process context diagrams are not detailed enough to give developers everything they need to implement a solution.

When you need to model processes in detail, it's best to use the standard that can satisfy those requirements. However, you need to keep in mind that mixing multiple standards in the same repository should be handled with extra care, as some elements and relationships may get duplicated, and some relationships between elements from different standards might not be possible. For example, you cannot link ArchiMate® elements to UML elements, but you can link UML elements to ArchiMate® elements.

It is possible to stop enforcing the standard relationships rule by choosing **Start** > **Desktop** > **Preferences** > **Preferences** > **Links** and unchecking the **Strict Connector Syntax** option, as shown in the following figure:

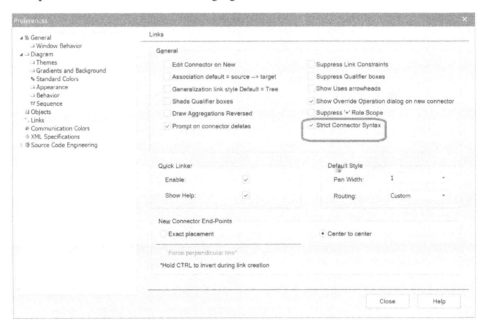

Figure 5.17 – The Sparx Preferences dialog

By disabling **Strict Connector Syntax**, Sparx will never enforce a syntax check on your connectors, which means that you can create any relationship between any two objects you choose, regardless of the standard. This way, you can create links between ArchiMate® and UML elements without any restriction. However, with flexibility comes more responsibility. Sparx will not be able to prevent you from creating incorrect relationships anymore if you disable this feature, so please be careful when you decide to opt for this. We do not advise you to do it unless you have a strong justification.

Since this book is about ArchiMate®, we will not go into more detail about creating UML diagrams. The main purpose of mentioning UML as an example is to let you know that your repository can consist of elements from multiple standards, even if this adds a level of complexity to managing and operating your repository. However, do not forget that enterprise architecture is about bridging gaps, not creating new ones. If stakeholders want processes to be modeled using UML, then let it be. It is easier for you to adapt than for the entire enterprise to adapt.

Another reason for mentioning UML is because we want to encourage developers and solution architects with no ArchiMate® experience to contribute to the repository with a standard that they feel comfortable with. UML is only one example, but there are dozens of modeling standards, and luckily, most of them are supported in Sparx. Adding a new standard to the repository must be approved by the enterprise architecture governing body to avoid ending up with a different standard for each person, which is far away from being a standard.

> **Important Note**
> Managing multiple standards in one repository can be hard, but managing multiple repositories in multiple modeling tools is much harder. You need to unify the architecture artifacts into a single repository as much as you can. Having diagrams following different standards in **Visio**, **Lucid**, and **Visual Paradigm**, for example, does not help toward this objective.

In the next section, we will introduce a new application behavioral element, which is the application function.

Introducing application functions

"An application function represents automated behavior that can be performed by an application component" (https://pubs.opengroup.org/architecture/ archimate3-doc/chap09.html#_Toc10045397).

Application functions are application internal behavior elements, so they share a lot of characteristics with application processes and application interactions. However, application functions do not involve sequences, and they represent an abstracted ability that a component can perform.

Confusion between application functions and application processes is very common, and so is confusion between application functions and application services. We will help in identifying the differences between the three element types by highlighting the major differences. The easiest to identify are the processes because they involve a sequence, while functions and services are abstracted and do not involve a sequence. On the other hand, services are externally exposed while functions and processes are internal-facing. The following table can help you memorize the differences:

The element	Internal-facing	External-facing
Has no sequence	Function	Service
Has a sequence	Process	Not possible

Table 5.1 – Differentiating between functions, services, and processes

You may think that application services are external-facing, but at the same time, they have a sequence of activities that enables them to provide or perform the service, which is not quite accurate. The sequence that shows how application services work is actually internal application processes realizing those services. Callers to a service do not know what the sequence is, so the sequence itself is not exposed to the enterprise.

Application components can perform many functions, but not all of them are necessarily exposed to the external enterprise. Let's clarify this with a simple example. We don't know how Microsoft Word works internally, so we as end users have no idea of what functions are performed by Word. We can only see and use what Word exposes externally to us, so we can only access its services. Mail merge is an example of an exposed service, but we cannot see or access the functions or the processes that are behind it. Application functions represent what the component architects and developers see, while application services represent what the component users and consumers see.

ArchiMate® provides two notations to model application functions, the rectangular and the borderless notations, as shown in the following diagram:

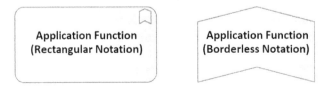

Figure 5.18 – Application Function notations

Application functions, processes, and services can have the same names, but that shouldn't confuse you either because it is all about the context. Printing, for example, is a function when we talk about an application's ability to print. Printing is considered a process when we start modeling how printing actually works, by defining the sequence of activities and the flow of data to be printed. Printing will be considered a service too if it is available for users to use. They all have the same name, but they are three different elements for three different purposes.

The following diagram shows the relationships between the three elements:

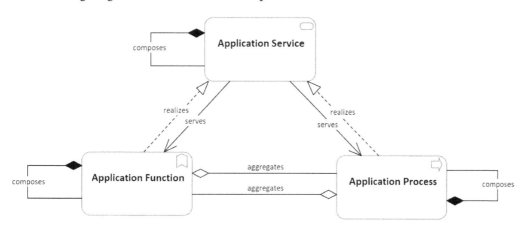

Figure 5.19 – Application functions, services, and processes

Note how an application function can be composed of application functions and aggregated of application processes. Application processes can also be composed of application functions and aggregated of application processes. This interchanged nested relationship can go to any desired level of depth until an atomic level of detail is reached, where elements cannot be broken down further. The following is an example of what the **Tracking App** application functions catalog can look like:

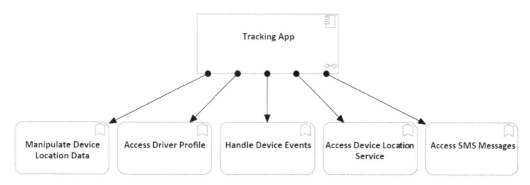

Figure 5.20 – The Tracking App application functions catalog

The focused metamodel for the application function is identical to the one for the application process, so we will not repeat it and leave it for you as an exercise. Just use *Figure 5.15* as a base metamodel and replace the application process element with the application function element.

Next, we will look at another application internal behavior element, which is the application interaction.

Introducing application interactions

"An application interaction represents a unit of collective application behavior performed by (a collaboration of) two or more application components" (https://pubs.opengroup. org/architecture/archimate3-doc/chap09.html#_Toc10045398).

Application interactions are application internal behavior elements, so they share a lot of characteristics with application processes and application functions. The only difference is that application interactions describe the behavior of a collaboration of components, while application processes and application functions describe the behavior of a single application component.

ArchiMate® provides two notations for modeling application interactions, the rectangular and the borderless, as you can see in the following diagram:

Figure 5.21 – Application Interaction notations

The focused metamodel of the application interactions should be identical to the one for application processes, so please refer to *Figure 5.15* and replace the focus element with the application interaction element.

Application interactions are not as commonly used as the other application internal behavior elements, so we will not spend too much time on them, and you can always refer to the ArchiMate® online documentation if you need to know more.

Next, we will talk about application events.

Introducing application events

"An application event represents an application state change" (https://pubs.opengroup. org/architecture/archimate3-doc/chap09.html#_Toc10045400).

A state change can occur due to internal or external factors such as a click on a command button by a user, reception of data from another application, completion of an application process, reaching a specific number of records, and reaching a predefined point of time. **Event-driven programming** is a very common way of building applications that react to events.

ArchiMate® provides two notations for modeling application events, the rectangular and the borderless, as shown in the following diagram:

Figure 5.22 – Application Event notations

Application events trigger the other application behavior elements, such as application processes, application functions, application services, and application interactions. Application events can also be triggered by those same elements. This means that an application event can trigger an application process to perform something specific, and when the process finishes, it can trigger another event, which in turn can trigger a second process, and so on. *Figure 5.16* shows how one event triggers the **Manipulate Device Location Data** application process and shows that a second event is triggered upon process completion.

Application events realize business events and are realized by technology events. Business events such as **Truck Moved** are like application events but have pure business meaning. An application event such as **Reception of Device Location Data** can realize the **Truck Moved** business event. Additionally, the **Reception of Device Location Data** application event can be realized by the **Phone GPS Location Changed** technology event, as shown in the following figure:

Figure 5.23 – Events realization at different layers

To build a reference that we can use in the future for modeling application events, we need to define a focused metamodel for it. At this point, we will assume that you have built some confidence in reading and understanding ArchiMate®'s metamodels and how to build a focused metamodel diagram in Sparx using your understanding. Therefore, we will not list the instructions for doing so and will provide the focused metamodel in the following diagram:

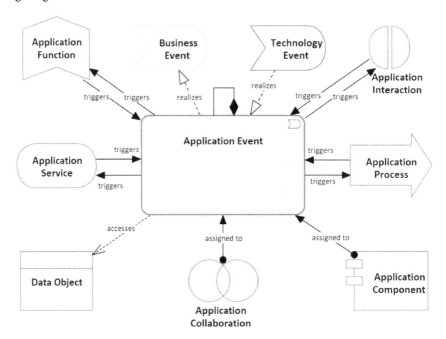

Figure 5.24 – The Application Event-focused metamodel

With this, we have reached the end of the second section of this chapter, which provided multiple examples for describing the behavior aspect of an application. In the next section, we will learn different ways of describing the structure of an application.

Describing application structure

The **application structure** describes the parts of the application that will be built. This includes the components and subcomponents, the interfaces, and the collaborations. They are the *tangible* parts of the application that can be deployed, copied, moved, deleted, or accessed. In this section, we will see how you can describe different parts of the application structure using different element types and relationships.

We will start this section by revisiting the application component, introducing the application interfaces, and then we will explore the application collaborations and how they can be helpful elements when modeling large applications.

Revisiting the application component

The Tracking App application component was introduced in *Chapter 3*, *Kick-Starting Your Enterprise Architecture Repository*, and the focused metamodel was introduced in *Chapter 4*, *Maintaining Quality and Consistency in the Repository*. So, we will not repeat what has been mentioned in the previous chapters, but we will introduce additional models that can be developed to give additional information about an application from different points of view.

It is important to bear in mind that as an enterprise architect, you do not usually enforce design decisions on the solution unless there are strong requirements that justify such enforcement. Your designs must stay at conceptual and logical levels where they can tell what is needed but not how to build it. What we will be doing in this subsection is looking at artifacts in the repository from the perspective of a solution architect and translating the conceptual ideas into logical designs. Solution architects are usually the ones who convert conceptual designs into logical and physical designs, and they are usually the ones who decide which design pattern and technologies to use.

When we introduced application functions, we modeled the Tracking App application functions catalog in *Figure 5.20*, but we didn't specify how these functions will be built. One solution architect may decide to build all these functions in a single component, a second solution architect may decide to build each function in its own subcomponent, while a third one may decide to follow a different design pattern.

For the sake of the context of this book, we will only show how to model the mapping between application functions and application subcomponents, without going into the details of which design pattern best fits our case. The target model will look like the following:

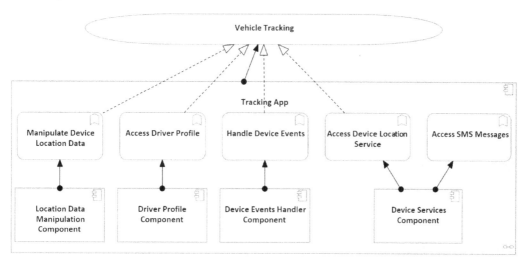

Figure 5.25 – Application functions mapped to application subcomponents

This model tells that the five desired functionalities will be implemented in four application subcomponents. Each component indicates which function or functions it is assigned to or, in other words, implements. Each of these subcomponents can be detailed even further, and more drill-down models can be developed until the *desired* level of detail is reached.

If the solution architect decided to go with a full-fledged microservices architecture, they may decide to have a separate application subcomponent assigned to each application process, no matter how small the process is. This gives ultimate scalability, extendibility, and reusability but adds a cost of development and maintainability overhead.

Another architect may decide to have all these functions realized in one monolith component for easier development but at the price of scalability, extendibility, and reusability. Deciding which design pattern to choose is not within the scope of this book. We have shown you one example for creating this mapping, and you can apply it in the way that best fits your environment and business requirements.

In the next subsection, we will introduce a new structure element, which is the application interface.

Introducing application interfaces

The **application interface** *"represents a point of access where application services are made available to a user, another application component, or a node"* (https://pubs.opengroup.org/architecture/archimate3-doc/chap09.html#_Toc10045394).

Application components encapsulate their internal structure and behavior and hide them from the external enterprise. For other enterprise elements (such as users, other application components, and system hardware and software), to access the services that are provided by an application component, they need to access them through an application interface. Application interfaces provide *predefined controlled* agreements known as **service contracts** to provide the services. Enterprise elements need to adhere to service contracts, or they will not be served. Service contracts provide the **data schema**, which describes the structure and format of data going into and out of an application interface. If you have experience in SOA, you must be familiar with modeling application interfaces, so you can easily apply your skills here.

ArchiMate® 3.1 provides two notations to model application interfaces, as you can see in the following figure:

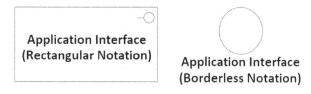

Figure 5.26 – ArchiMate® 3.1 Application Interface notations

As always, it is up to you to choose the notation that creates less confusion for you and, more importantly, your audience. Let's see how the application interface-focused metamodel looks before looking at some modeling examples.

Defining the application interface-focused metamodel

As we have done for other focused metamodels, we will use the ArchiMate® 3.1 specification to guide our focused metamodel development. We will translate our understanding and interpretation of the standard into an easy-to-understand reference that will help to maintain consistency within the enterprise architecture repository.

Since you already have confidence in reading the ArchiMate® metamodels and building focused metamodels in Sparx, we will skip the step-by-step instructions and introduce the focused metamodel directly. Your application interface-focused metamodel should look like the following:

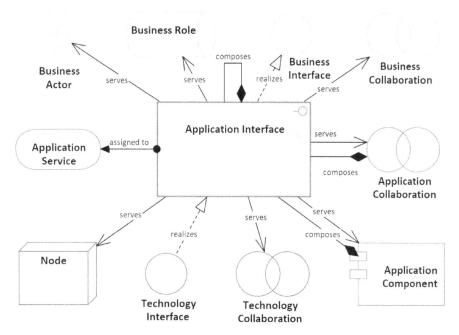

Figure 5.27 – The Application Interface-focused metamodel

We hope that this focused metamodel will guide you in modeling more application interfaces in your work environment and empower your team with clearer and easier-to-follow references.

In the next subsection, we will look at the different types of application interfaces and ways of modeling them.

Modeling application interfaces

As we have mentioned in this section's introduction, **application interfaces** are the access point for accessing application services. *Figure 5.7* has already covered this mapping by showing how the service will be accessible through six different application interfaces. What might be helpful to show now is how these interfaces will be structured within the **Tracking App** component.

Even though we will not enforce any design decisions, we know for sure that we will need at least two UIs – one for Android users and one for iOS users. We will most probably need one for web users, one for the command line, and one for API calls. However, application interfaces are not deployable objects. They cannot exist by themselves and can only be part of an application component. This means that to have an Android UI, you need to have an Android UI component that will be installed on the phone. That component will contain the Android UI and be responsible for sending and receiving data from and to the UI. The same logic applies to the iOS application interface.

The following model shows how **Tracking App** can be composed of multiple UI application components, and multiple application interfaces as well:

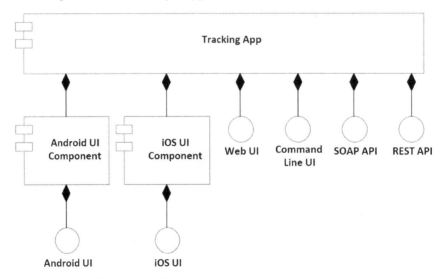

Figure 5.28 – The Tracking App application interfaces

You may agree or disagree on the design, but remember that we're just clarifying the idea of modeling application interfaces with examples. Another architect may decide to have the web UI, command-line UI, SOAP API, and REST API each in a separate application component, just like the Android UI component, which are valid to use. Using different design patterns and the advantages of one over another are covered in many other books on the market. From a modeling perspective, we hope that you've got the idea that you can apply the idea in any way that makes sense to you and is acceptable to your team.

If you want to detail how the command-line UI works, for example, you can model that in a separate child diagram, as follows:

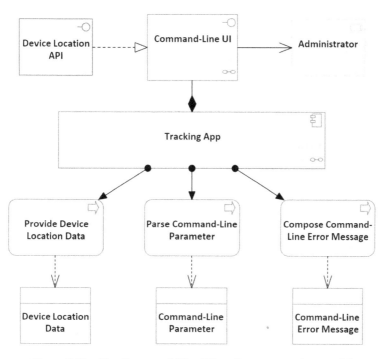

Figure 5.29 – The Command-Line UI application interface model

As we said earlier, you may agree or disagree with the design itself, but this is one possible way to model an application interface in ArchiMate®. Once you get the idea and understand the local issue that you need to resolve, the sky is your limit to model whatever makes sense in your work environment.

In the next section, we will introduce another application structural element, which is application collaboration.

Introducing application collaborations

"An application collaboration represents an aggregate of two or more application internal active structure elements that work together to perform collective application behavior" (https://pubs.opengroup.org/architecture/archimate3-doc/chap09. html#_Toc10045393).

Application collaborations, according to the definition, are structural elements that are aggregated of two or more structural elements. A very common example is Microsoft Office, which is an aggregation of Word, Excel, PowerPoint, Outlook, and other products. Each product is independent of the other, but combined, they are known as the Microsoft Office product. This is an example of weak collaboration where individual components still maintain their independence.

Some application collaborations can take stronger forms and are more than just a gathering of multiple components, and more like unification. When independent components are combined, they provide a new set of application behaviors that cannot be performed by any one of them separately. Additionally, in some cases of strong application collaborations, access to the individual components will no longer be allowed through their own interfaces but through new ones that belong to the collaboration element. It all depends on how strong or weak you or the solution architect want this collaboration to be.

ArchiMate® 3.1 provides two notations to model application collaboration elements, the rectangular and the borderless, as with most of the ArchiMate® elements:

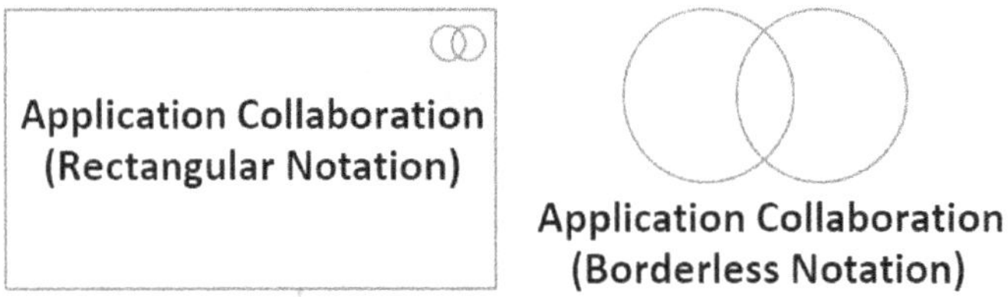

Figure 5.30 – Application Collaboration notations

The **Application Collaboration** element has the same relationships that the application component has, so they both share the same metamodel. Refer to *Figure 4.9* in *Chapter 4*, *Maintaining Quality and Consistency in the Repository*, to see the possible elements that can relate to an application collaboration and the relationships that can occur between them.

Figure 5.31 is an example of an application collaboration. It shows a conceptual design of the e-commerce platform that *ABC Trading* is targeting to build, which aggregates multiple application components and provides multiple application interfaces. It also shows that the platform will do the following:

- Be served by cloud computing services
- Be served by the single sign-on service
- Provide an e-trading service
- Provide users with the ability to manage a unified user profile for all these different components

One thing to keep in mind about application collaborations is that they can assign their internal application behavior to application interaction elements in addition to application processes and application functions, as shown here:

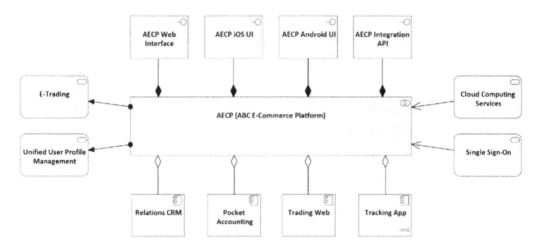

Figure 5.31 – An AECP application collaboration model

That was an example to simplify the idea of modeling application collaborations, and we're quite sure that you will be able to project this example onto real work examples to provide your stakeholders with models describing their visionary solutions. That's what enterprise architects do – help stakeholders have better views from different perspectives to enable them to make better decisions.

In the next section, we will see how to model data in the enterprise without *stepping on the toes* of database administrators.

Introducing data objects

"A data object represents data structured for automated processing" (https://pubs.
opengroup.org/architecture/archimate3-doc/chap09.html#_
Toc10045404).

Applications are built to automate the process of manipulating data in one way or
another, whether it is financial data, human resources data, inventory data, video clips,
soundtracks, virtual reality images, games, and so on. The list is long.

Data, by itself, is of little use if it is not opened and processed with the right application
component. A .jpg file, for example, does not have any value for you if you don't have
a JPG viewer installed. A database record has no value to you if it is not displayed on the
screen, processed, reported in a report, printed on paper, or sent to another application
component. Therefore, ArchiMate® 3.1 considers data objects as passive elements because
they must be manipulated or *accessed by* another element.

Unlike most elements in the ArchiMate® 3.1 specification, there is a single notation for
modeling data objects, as shown in the following figure:

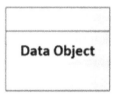

Figure 5.32 – ArchiMate® 3.1 Data Object notation

Data objects must not be confused with databases. Databases are one type of storage for
data objects. Data objects can exist in flat files, spreadsheets, relational tables, documents,
and many other forms. The driver profile in our Tracking App, for example, is a data
object, regardless of where it is stored and in what format. Whenever we describe the
actual locations where the driver profile data is stored or how it is physically persisted,
then we're more likely describing the technology object that realizes this data object.

Let's talk more about data at the different layers of the enterprise and how to differentiate
between them.

Differentiating data at the different architecture layers

At the technology layer, we can model how data is stored in the form of technology objects and technology artifacts. You can think of technology objects as **data at rest**, while data objects represent **data in motion or transit**. A high-level technology object model can show the storage of the **Driver Profile as a Profiles Data store**.

Data stores do not tell us whether we are storing data in a relational database, a `.json` file, or a flat text file. All of these are considered technology artifacts realizing data objects from the application layer. When you need to provide a deeper view of this data store, such as the database tables that form it, then you are moving away from being abstract to being more specific, which is another view for another stakeholder.

A database table is yet another technology object at a lower level of detail. A third level of detail can describe the data schema for a specific `.json` file with all the required tags and headers to physically represent the file. At this level of detail, ArchiMate® provides another element, the technology artifact, which is a specialized element of technology objects. We will look at technology objects and technology artifacts in more detail in *Chapter 6, Modeling in the Technology Layer*.

Data objects at the application layer represent data instances in memory. Data objects show the movement of data between components. As you can see in *Figure 3.16* of *Chapter 3, Kick-Starting Your Enterprise Architecture Repository*, and *Figure 5.12*, we have indicated the data objects that were passed in or out of different types of elements. Data objects realize **business objects**.

Modeling data at the business layer does not show any physical or implementation detail. The business might treat all employee profiles the same way, and it does not matter to the business whether these profiles are in a computer system or a paper folder. All that matters to the business is that there is a data structure that holds employees' profiles, including drivers' profiles. Business objects are realized by data objects when they are implemented within an application. We will talk in more detail about business objects in *Chapter 8, Business Architecture Models*.

Next, we will look at our final reference-focused metamodel in the application layer. More elements will be introduced in later chapters, and more focused metamodels will be built.

Defining the data object-focused metamodel

If you have read the ArchiMate® specification properly and modeled it in Sparx the way that we have guided you, your **Data Object**-focused metamodel should look like the following:

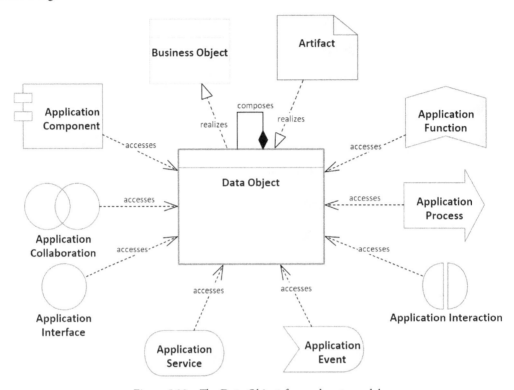

Figure 5.33 – The Data Object-focused metamodel

Let's look at a data object model to see how data objects can be modeled.

Modeling data objects

There is no specified size for how big or small a data object can be. It can be as small as a single attribute that you pass from one application process to another. It can also be as big as a dataset that composes many objects into one big object.

The **Driver Profile** data object, for example, can consist of many smaller data objects, such as user accounts, driver demographics, addresses, and contact numbers. Each of them can be decomposed further into smaller data objects, and they can also be aggregated with other data objects to form new data objects:

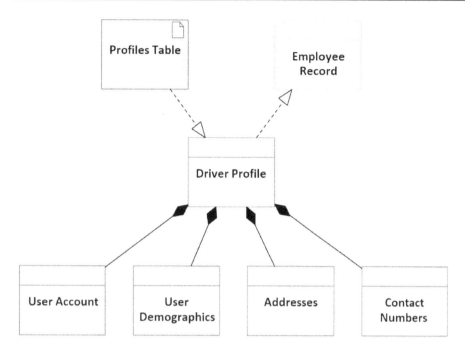

Figure 5.34 – The Driver Profile data object decomposition

In some cases, you may find a need to pass the entire driver profile between components, while in some other cases, you may only need to pass an address. The question that will come to the architect's mind is, *shall I define the attributes of the data objects?*

The answer is, it depends on your target audience, but in most cases, you do not have to define the attributes unless you are passing attributes individually between components and want your model to clearly depict this data exchange. If you are defining the attributes, remember that you need to coordinate with the data architect to ensure that you are not forcing any design decision on them.

There are many ways to model data objects, but to create consistent models that adhere to standards, you will need a metamodel. Let's recap what we have learned in this chapter.

Summary

In this chapter, we've covered some of the most fundamental aspects of enterprise architecture modeling. We've looked at the differences between behavioral and structural models and the elements that make them up. We've also looked at some common diagram types, such as the context diagram and the catalog. We've learned more about how to manipulate our models in Sparx and how to connect them to each other to provide an easy navigation experience for readers.

You will see that many of these aspects will be reinforced in subsequent chapters as we cover aspects of the technology and business layers of ArchiMate®. This is because many of the element types across these layers share a common function but with a differing scope. There are services, functions, processes, events, components, and interfaces in the business, application, and technology layers. If you understand the thin lines that differentiate each element in each layer, you've mastered ArchiMate® and enterprise architecture, literally!

We will continue to use the focused metamodel in introducing new elements; however, we are guessing that you've had enough practice creating your own focused metamodels and that we don't need to show you how to do that anymore. We'll just introduce the models. We will also continue to salt each chapter with tidbits of advice from our decades of experience. Feel free to take that advice with a grain of salt. This practice we call architecture is always changing. You need to change with it.

In the next chapter, we will cover the technology layer of the enterprise.

6
Modeling in the Technology Layer

One of the difficult tasks enterprise architects must tackle is describing technology environments at the correct level of abstraction. If the model is too detailed, as architects, we are seen as too prescriptive. If it's not detailed enough, we are being vague. Either of these conditions tends to lead to the **ivory tower syndrome** as we are seen as being too far from reality. Selecting the perfect level of granularity can sometimes seem like an art.

The creators of the ArchiMate® language recognized this fact and took it into consideration when designing the language. Each layer's elements are based on a small, core set of elements that have similar attributes across layers (see *Figure 6.13* further on in this chapter). As you read through this chapter, you should see these similarities begin to appear. You should also begin to get a sense of the level of granularity necessary for architecture descriptions and how to create them without duplicating information.

In this chapter, we cover elements and relationships of the **Technology Layer** of the **ArchiMate® 3.1** standard. This includes a new sub-layer of the Technology Layer called the **Physical Layer** for modeling physical environments. The following sections are included in this chapter:

- Modeling technology environments
- Modeling physical environments
- Modeling networks

As always, we include examples and tips for using **Sparx** along the way.

Technical requirements

This chapter does not require any additional technical requirements other than having Sparx Systems **Enterprise Architect** (**EA**). If you do not have a licensed copy, you can download a fully functional 30-days trial version from the Sparx Systems website (`https://sparxsystems.com/products/ea/trial/request.html`).

We will continue adding the content of this chapter in the same EA repository that we built in *Chapter 3*, *Kick-Starting Your Enterprise Architecture Repository*, and *Chapter 4*, *Maintaining Quality and Consistency in the Repository*. If you have not read these two chapters, we strongly advise you to read them first then come back to read this chapter. If you want, you can download the repository file of this chapter from GitHub at `https://github.com/PacktPublishing/Practical-Model-Driven-Enterprise-Architecture/blob/main/Chapter06/EA%20Repository.eapx` instead of starting from scratch. Some of the steps in this chapter depend on elements that have been already created in the repository, so it is better not to start this chapter with an empty repository.

We will use the following ArchiMate® 3.1 specification chapters to guide our development:

- *Chapter 5*, *Relationships* (`https://pubs.opengroup.org/architecture/archimate3-doc/chap05.html#_Toc10045310`)
- *Chapter 10*, *Technology Layer* (`https://pubs.opengroup.org/architecture/archimate3-doc/chap10.html#_Toc10045407`)
- *Chapter 11*, *Physical Elements* (`https://pubs.opengroup.org/architecture/archimate3-doc/chap11.html#_Toc10045429`)
- *Chapter 12*, *Relationships Between Core Layers* (`https://pubs.opengroup.org/architecture/archimate3-doc/chap12.html#_Toc10045440`)

Modeling technology environments

In this section, we will cover the core elements of the Technology Layer of ArchiMate® 3.1. The elements discussed in this section are identified in the following diagram:

Figure 6.1 – Technology structural elements

At the core of the Technology Layer of ArchiMate® are two new elements: the **Device** and **System Software** elements. These two elements are both technology internal active structure elements and they both inherit from the same parent element, the technology node. As per the ArchiMate® 3.1 specification, a technology node—or **node**, in short— "*represents a computational or physical resource that hosts, manipulates, or interacts with other computational or physical resources*" (`https://pubs.opengroup.org/ architecture/archimate3-doc/chap10.html#_Toc10045410`).

In simpler words, a node is any hardware or software that is used to store, process, monitor, manage, streamline, and communicate with other hardware or software. If your background is in networking, you may be wondering about network nodes. ArchiMate® makes no graphical distinction between network nodes and other types of nodes. For this reason, we usually refine the node type by giving it an appropriate name. For example, a network node may actually contain the word *network* in the element name.

As with most enterprise elements, ArchiMate® provides two notations for modeling nodes—rectangular notation and borderless notation, as you can see in the following screenshot:

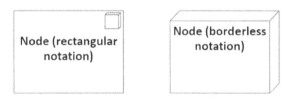

Figure 6.2 – ArchiMate® 3.1 node notations

Since the device and the system software elements are a specialization of the node element, they have the same relationships as the node. This means that all the relationships that you can see in the technology node-focused metamodel (see *Figure 6.12*) are applicable for the device and system software. There are cases where you need to be more precise about what a specific node is, and this is where you need to use either a device or system software. Let's take a look at some examples of modeling with node elements.

Examples of technology models

We need to model some of the technology components involved in serving web requests at *ABC Trading*. For this, we will build up a diagram incrementally as we go. This approach lets us demonstrate a feature of Sparx, the diagram layering feature. We'll explain that later. First, let's look at the diagram here:

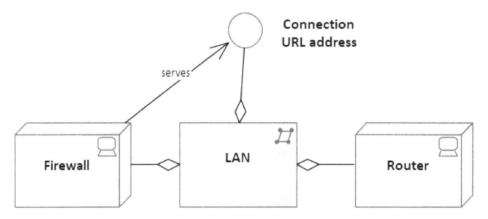

Figure 6.3 – Layer 1 network technologies

We've not covered these element types in detail yet, but we will. The device element will be covered in the *Using the device element* subsection. The network element is presented in the *Modeling networks* section of this chapter. For now, let's walk through the first iteration of this diagram. We see that the following applies:

- We have a **Local Area Network** (**LAN**).

- We use the aggregation relationship to show that the LAN is made up of a **firewall** and a **router**.

- We interface with the LAN through a **connection Uniform Resource Locator (URL) address**.

- The firewall serves to protect the interface. We go into a little more detail on this in *Chapter 7, Enterprise-Level Technology Architecture Models*.

While that's not much information, we will build upon it. The next layer shows some of the devices that connect to the LAN, as illustrated in the following diagram:

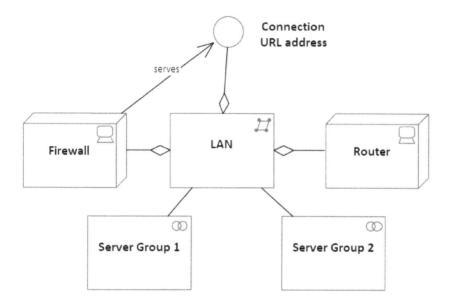

Figure 6.4 – Layer 2 connected server groups

In this layer, we have added server group elements. We've used the collaboration element type to represent a server group. In this case, we have a collaboration of servers that serve web pages. We see that there are two server groups connected to the LAN. Unlike the firewall and router, the server groups do not make up the LAN—they connect to it. We show a connection using a simple *association* link. The next layer expands on the server groups, as illustrated in the following diagram:

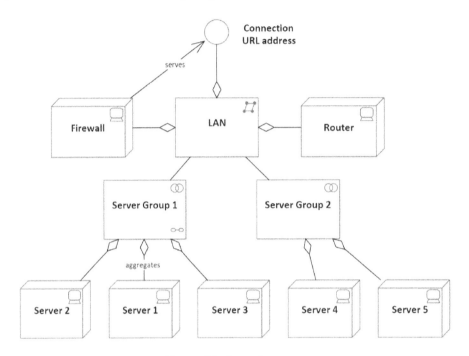

Figure 6.5 – Layer 3 servers

In this layer, we see that **Server Group 1** is made up of three servers, and **Server Group 2** has two servers. The next layer shows that all these servers serve the same set of web pages, as illustrated in the following diagram:

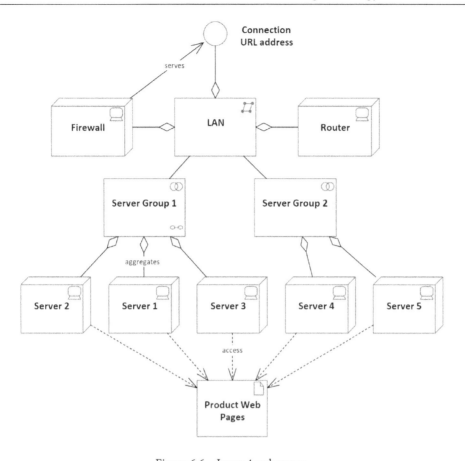

Figure 6.6 – Layer 4 web pages

There are two problems with Layer 4 in the diagram. As we know, servers don't serve web pages directly. We're missing some intermediate technologies.

The other problem is that this diagram is already becoming a bit too cluttered. We need to pare it down so that the reader can focus on what we are conveying. We've all seen the huge **Entity-Relationship Diagrams** (**ERDs**) with 50 or more elements and endless connectors running all over the place. They often are hung on the wall like a trophy as if to say: *Look what I have conquered!* While they may look impressive or intimidating, they are usually very ineffective. The information may be useful, but the diagrams are useless— they are too difficult to read. We also need to consider our audience. How much of what is presented in *Figure 6.6* is valuable to them? They probably don't need to know about the individual servers involved in the solution.

So, how do we add the missing technologies without overloading our diagram or our audience? The answer is to create another diagram. Simply select an area of the diagram that needs more detail and make it the focus of the next diagram. In the following diagram, we have expanded the information on the **Server Group 1** collaboration element:

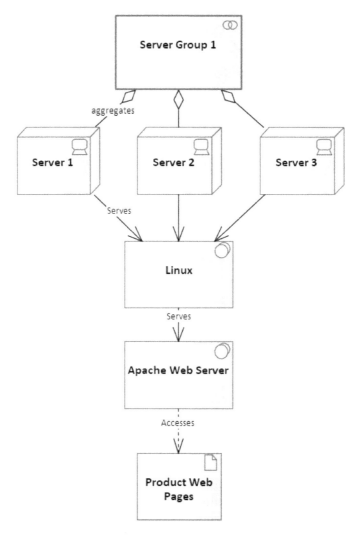

Figure 6.7 – Server group 1 detail

When creating diagrams that are dependent on each other such as we have just done, we relate the diagrams by including a common element. In our case, the element that links our two diagrams is the server group 1 element. All of the rest of the elements should be unique to each diagram. If you look back at *Figure 6.5*, *Figure 6.6*, and *Figure 6.7*, you'll notice that we have included server elements in all of them. We need to remove the servers from *Figure 6.6*. You might think: *Why bother?* But should we ever need to change the servers that comprise server group 1, we would prefer to minimize the number of diagrams we need to modify. The following diagram shows the new parent technology stack:

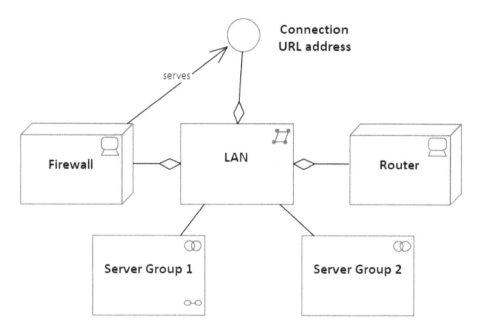

Figure 6.8 – Final web-servicing model

Additionally, we added *Figure 6.7* as a child diagram to the server group 1 element in *Figure 6.8*. This is indicated by the chain-link icon in the lower-right corner of the element. Now, we can easily navigate between the two diagrams. We have reduced clutter on each diagram, but we have also provided two layers of granularity for our audience. You can now choose the level of detail provided to your audience without duplicating information. Next, we will cover the diagram-layering feature of Sparx that we used to build and display this structure diagram.

Diagram filters and layers

Adding filters or layers to a diagram in Sparx is a very convenient way to show diagram features and elements incrementally. The **Filters** option allows us to hide or mask elements based on attributes such as stereotype, version, status, and date. Filters are diagram-independent—they can be applied to any diagram. The **Layers** option performs a similar function based on the diagram elements you select. For that reason, layers are specific to a diagram. To access this feature, go to **Layout** > **Tools** > **Filters & Layers**. The following window appears:

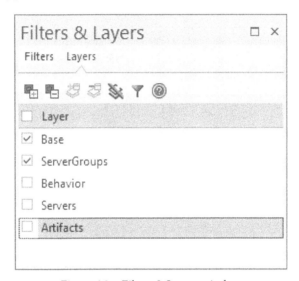

Figure 6.9 – Filters & Layers window

As with other windows in Sparx, you can choose to dock this window on the Sparx **User Interface** (**UI**). The first two buttons on the **Layers** ribbon allow you to add or remove a layer from a diagram. The next two buttons let you add or remove selected elements from a selected layer. Finally, checking or unchecking the checkbox beside **Layer** displays or hides elements of a layer from a diagram. Just as with any other feature, it's best to experiment with this one to fully understand how it works.

If you'd like to play around with this example diagram, you can find it in this chapter's GitHub repository at `https://github.com/PacktPublishing/Practical-Model-Driven-Enterprise-Architecture/blob/main/Chapter06/EA%20Repository.eapx`. Once you're in the Sparx repository, the diagram can be found in the following location:

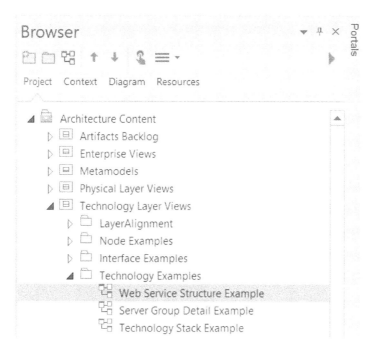

Figure 6.10 – Example diagram location

Now, let's look at another way to model system structure—the technology stack.

Technology stacks

One specific technique for modeling system structure is a **technology stack** diagram. To our knowledge, there is no standard for this technique; it's more of a common practice. The idea is to represent technology dependencies by placing dependent technologies on top of the technology on which it is dependent. In this sense, we literally *stack* the technologies. Have a look at the following example depiction of a **Linux-Apache-MySQL-PHP: Hypertext Preprocessor (PHP) (LAMP)** technology stack:

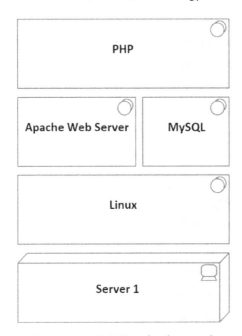

Figure 6.11 – LAMP technology stack

LAMP represents one somewhat common combination of technologies used for web development. There are many others, of course.

In this diagram, each system software element is dependent on the one below it. Dependencies flow down, never up. So, in this example, PHP is dependent on all of the other technologies. As you can see, there are no visible links in this diagram. They are all implied. This renders such diagrams of somewhat limited use. There is no information conveyed by technology stack diagrams that cannot be conveyed in other standard structure diagrams. We include it here because if you've never heard of it before, you will likely hear of it in the future.

Interpreting the standard

To build a node-focused metamodel, we will need to interpret the standard into a list of statements describing elements and relations; then, we will translate these statements into a diagram. We have done similar steps for many elements in previous chapters, so if you have not read *Chapter 4, Maintaining Quality and Consistency in the Repository*, we highly recommend that you do because it contains step-by-step instructions for performing many of the actions in this chapter and most of the remaining chapters.

A node is a technology element, so we will use the following chapters from the ArchiMate® 3.1 specification as references:

- *Chapter 5, Relationships* (`https://pubs.opengroup.org/architecture/ archimate3-doc/chap05.html#_Toc10045310`)

- *Chapter 10, Technology Layer* (`https://pubs.opengroup.org/ architecture/archimate3-doc/chap10.html#_Toc10045407`)

- *Chapter 12, Relationships Between Core Layers* (`https://pubs.opengroup. org/architecture/archimate3-doc/chap12.html#_Toc10045440`)

We have all the information that we need to build a focused metamodel, so we will put all the aforementioned points in a diagram in Sparx, as you will see in the next subsection.

Using the node element

Since you have more confidence and experience in Sparx by now, we will assume that you know where to find all these elements, how to change toolboxes, how to place elements, how to create proper relationships, and how to style your elements and organize your diagram. Whenever you're in doubt, refer to *Chapter 4*, *Maintaining Quality and Consistency in the Repository*. The node-focused metamodel should look like this:

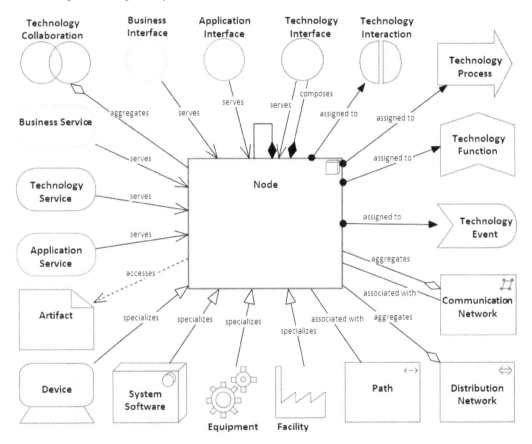

Figure 6.12 – Technology node-focused metamodel

This diagram references the core elements of the Technology Layer. We will add some more elements and relationships in the *Modeling physical environments* section later in this chapter.

After reading the previous chapter, you may have detected a pattern forming among the layers of ArchiMate®. You are not mistaken. It is no accident that the elements in each layer have complementary elements in the other layers. These complementary elements have similar characteristics within the scope of their layer. For example, the business actor element performs a similar function in the Business Layer, as does the component element in the Application Layer and as the node element in the Technology Layer. The following diagram shows the alignment of most of the core elements in ArchiMate® 3.1:

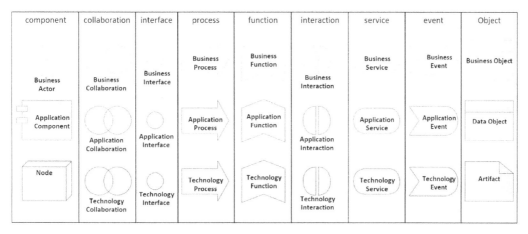

Figure 6.13 – Element alignment among layers

Now that we've covered nodes, we take a deeper look at some of the specific structural elements in the Technology Layer. In this section, we will present focused metamodels for the following structural elements:

- Device
- System software
- Technology interface
- Technology collaboration

Network and **path** element types are covered in the *Modeling networks* section later in this chapter. **Technology behavioral** elements are covered in *Chapter 7, Enterprise-Level Technology Architecture Models*. Let's get started.

Using the device element

According to ArchiMate®, "*A device represents a physical IT resource upon which system software and artifacts may be stored or deployed for execution.*" (`https://pubs.opengroup.org/architecture/archimate3-doc/chap10.html#_Toc10045411`)

The device element icon looks like a terminal or desktop computer, as illustrated in the following diagram:

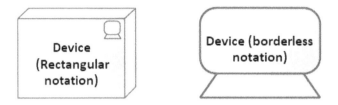

Figure 6.14 – Device icon

The device element is a specialization of the node element. It represents a physical device or piece of equipment. Examples of devices include:

- Servers
- Monitors
- Routers
- Cell phones
- Tablet computers
- Mainframe computers

Any physical **Information Technology** (**IT**)-related thing that may serve the Application Layer or other elements in the Technology Layer can be a device. In the following diagram, we show that a **Desktop PC** is an aggregation of a **CPU**, **Keyboard**, **Monitor**, **Mouse**, and **Cables**:

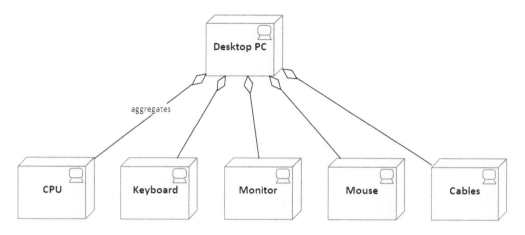

Figure 6.15 – Device example

Because a device is a type of node element, its relationships look similar to those of a node, as illustrated in the following diagram:

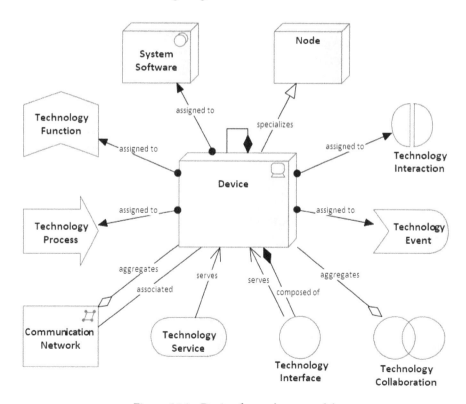

Figure 6.16 – Device-focused metamodel

Devices can be made up of other devices. Devices may be assigned to system software, technology functions, processes, interactions, or events. Devices may collaborate with other devices to provide functionality.

Where a device exists, there is often system software to manipulate that device. Let's look at the system software element next.

Using the system software element

According to ArchiMate®, "*System software represents software that provides or contributes to an environment for storing, executing, and using software or data deployed within it.*" (`https://pubs.opengroup.org/architecture/archimate3-doc/chap10.html#_Toc10045412`)

The system software icon is a box with a disk in the upper-right corner, as shown in the following diagram:

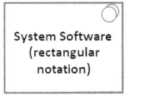 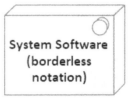

Figure 6.17 – System software icon

The system software element is a specialization of the node element. It represents software used to support the function of a device or other Technology Layer elements. Examples of system software include the following:

- Operating systems
- Application servers, web servers, and transaction monitors
- **Relational Database Management Systems (RDBMSs)**
- Backup and restore software
- Network firewall, routing, and monitoring software
- Report formatting systems
- Editors and word processors
- Security management software
- Device controller software
- Code compilers and interpreters

System software contains no business-specific logic. Once the software is *applied* to a business domain such as finance or inventory, it becomes an **application**. Application software can use system software to accomplish its business goal, but the system software knows nothing about that goal. The best example of this is when application software runs on an operating system. In such a case, the operating system has no application-specific knowledge. It simply acts at the behest of the application to open files, allocate memory, and other such technical tasks.

The following diagram expands on the example presented previously to include system software such as operating systems:

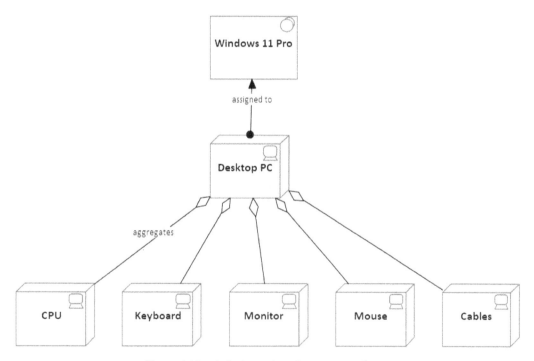

Figure 6.18 – A device assigned to system software

In this example, we see that a **Desktop PC** device is assigned to the Windows 11 Pro system software. In the next example, we see that system software may serve other system software:

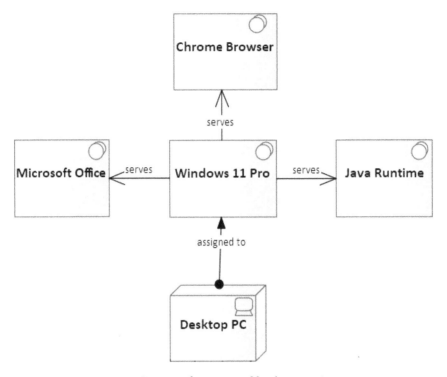

Figure 6.19 – System software served by the operating system

As with the device element, because the system software element is a type of node element, its relationships look similar to those of a node, as shown next:

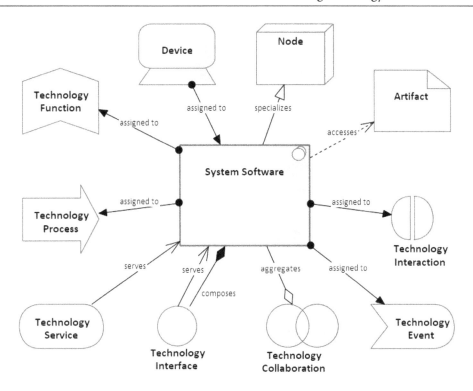

Figure 6.20 – System software-focused metamodel

Applications interact with system software through well-known technology interfaces. That's our next topic.

Using the technology interface element

The technology interface performs a similar function to the application interface, only in the Technology Layer. According to ArchiMate®, "*A technology interface specifies how the technology services of a node can be accessed by other nodes. A technology interface exposes a technology service to the environment.*" (https://pubs.opengroup.org/architecture/archimate3-doc/chap10.html#_Toc10045414)

The technology interface element icon shown next is a simple circle. Sometimes, when a connector is applied to it, it's referred to as a *lollipop*:

Figure 6.21 – Technology interface icon

The technology interface represents functionality exposed to its environment. Technology interfaces come in different forms. Some examples of technology interface types include the following:

- UIs

- **Application Programming Interfaces** (**APIs**)

- Communication interfaces (signals, sockets, web)

- File-based interfaces

The technology interface element can represent different levels of granularity or abstraction from generalized groups of interfaces such as the prior examples or to specific actions such as the following:

- Opening a file

- Creating a table

- Opening a socket

- Receiving a message

One example of system software that supports a rich set of interfaces is the **Linux** operating system. The following diagram depicts a few of those interfaces:

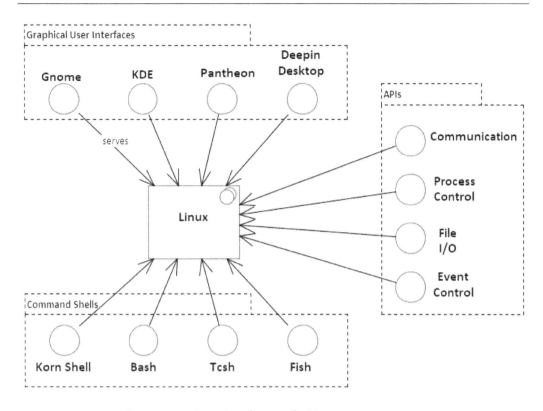

Figure 6.22 – Some interfaces on the Linux operating system

As you can see, Linux users can choose one of several **Graphical User Interfaces** (**GUIs**) or command shells to interact with the operating system. Linux also supports a rich set of APIs.

The most common relationships of the technology interface element are represented in the following focused metamodel:

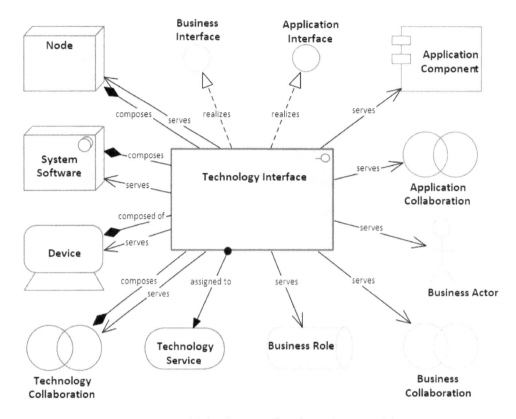

Figure 6.23 – Technology interface-focused metamodel

While an interface is generally exposed by a node, the functionality exposed by that interface may be the work of two or more nodes. Such functionality is a technology collaboration. We will look at the technology collaboration element next.

Using the technology collaboration element

According to the ArchiMate® standard, "*A technology collaboration represents an aggregate of two or more technology internal active structure elements that work together to perform collective technology behavior.*" (https://pubs.opengroup.org/architecture/archimate3-doc/chap10.html#_Toc10045413)

Just as with an application collaboration, the icon representing a technology collaboration appears as two intersecting circles, as shown here:

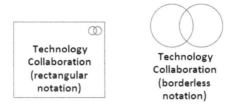

Figure 6.24 – Technology collaboration icon

In *Figure 6.5*, *Figure 6.6*, *Figure 6.7*, *Figure 6.8*, and *Figure 6.9*, we used a technology collaboration to represent a cluster of servers called server group 1 and server group 2. The servers in each cluster collaborated to deliver web pages. Technology collaboration elements are generally paired with technology interaction behavioral elements to describe the role of each element in delivering the functionality. Our example didn't need a technology interaction element because each server's role is identical to the next, and serving web pages follows a well-known interaction pattern. Besides, our focus in this chapter is on structural elements. We'll be reviewing behavioral elements in the next chapter.

The following diagram shows the relationships of the technology collaboration element:

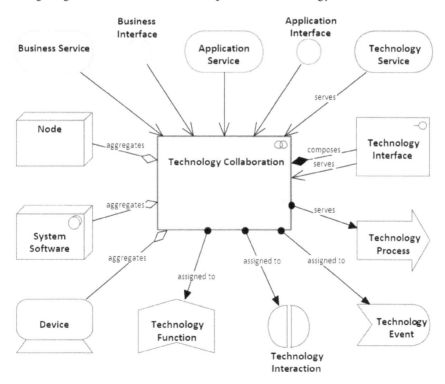

Figure 6.25 – Technology collaboration-focused metamodel

In the previous chapter, we learned about the **application data object**. The physical form of data in the Technology Layer is an **artifact**. We discuss that element next.

Using the technology artifact element

In ArchiMate®, an artifact is to the Technology Layer what a data object is to the Application Layer and what a business object is to the Business Layer. That is to say that an artifact represents data but in physical form. Unlike the Technology Layer elements discussed so far, an artifact is a specialization of a technology object (`https://pubs.opengroup.org/architecture/archimate3-doc/chap10.html#_Toc10045426`).

The technology artifact element resembles a box with an icon in the corner that resembles a piece of paper with one corner folded over, as shown here:

Figure 6.26 – The technology artifact

Examples of technology artifacts include the following:

- Physical data files
- Deployable objects such as `.jar` files and `.war` files
- Database tables
- Source files, executable files, and scripts
- Configuration files

The name of the artifact element should reflect the name of the physical file. Let's take a look at this element's focused metamodel in the following diagram:

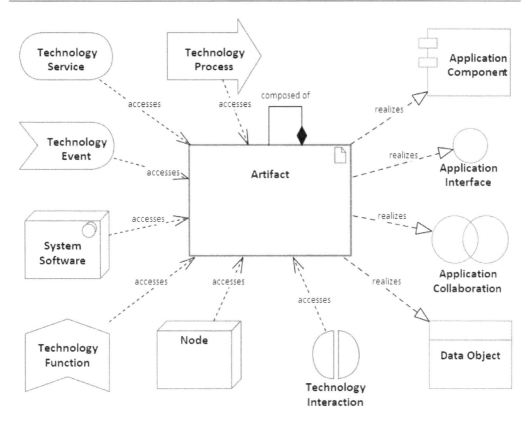

Figure 6.27 – Artifact-focused metamodel

As mentioned earlier, there are other structural elements in the Technology Layer, but we will look at those in the *Modeling networks* section later in this chapter. In the next section, we will look at modeling physical environments.

Modeling physical environments

Now that we understand the core Technology Layer elements in ArchiMate®, we need to establish a bit of context with respect to *ABC Trading*. Establishing context around a particular business domain can mean different things to different people. For me, once I have learned what a company does, I like to know where they do it, and with which resources.

In this section, we will use the elements and relationships of the physical sub-layer of the Technology Layer. The physical layer subset is relatively new to ArchiMate®. It was introduced in **version 3.1**. We will first look at the specific elements and relationships that make up the physical layer, then we will look at some examples of how those elements can be applied to *ABC Trading*.

The element types covered in this section are shown in the following diagram:

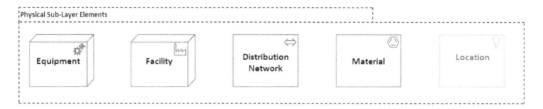

Figure 6.28 – Physical elements covered in this section

Let's have a look at each of these physical elements in the following subsections.

Understanding the equipment element

The **equipment** element serves as the core element of the physical layer (https://pubs.opengroup.org/architecture/archimate3-doc/chap11.html#_Toc10045432).

The equipment element is a specialization of the node element. It represents machines, tools, or other such physical mechanisms used by the organization. Equipment can comprise other equipment elements. The equipment icon resembles a set of meshed gears, as shown here:

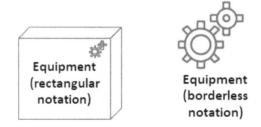

Figure 6.29 – Equipment icon

The following equipment-focused metamodel diagram shows some of the possible relationships of the equipment element:

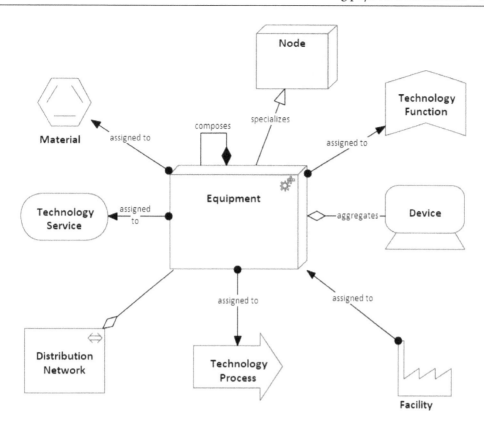

Figure 6.30 – Equipment-focused metamodel

Let's now turn to another structural element—the facility element.

Understanding the facility element

The **facility** element is a specialization of the node element and can be used to represent any physical structure or environment that plays an important role in the organization (https://pubs.opengroup.org/architecture/archimate3-doc/chap11. html#_Toc10045433).

The icon for the facility element resembles a factory, as seen here:

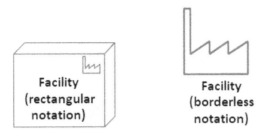

Figure 6.31 – Facility icon

The following facility-focused metamodel diagram depicts the relationships of the facility element:

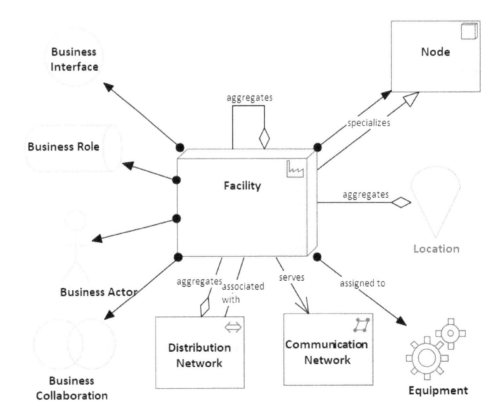

Figure 6.32 – Facility-focused metamodel

The next element in the physical layer is the distribution network element.

Understanding the distribution network element

The **distribution network** element is an object element. It represents a means of transporting physical things such as material, energy, and much more (https://pubs.opengroup.org/architecture/archimate3-doc/chap11.html#_Toc10045434).

The distribution network icon appears as a double-ended arrow, as depicted in the following diagram:

Figure 6.33 – Distribution network

The distribution network element can have the following relationships:

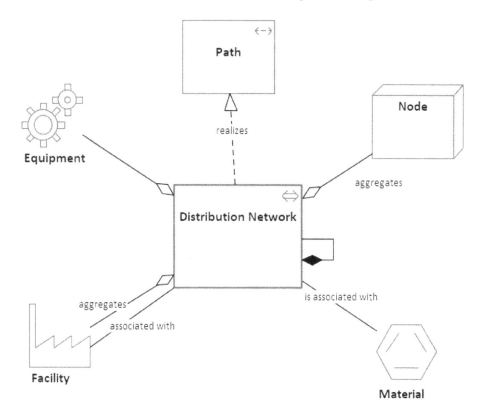

Figure 6.34 – Distribution network-focused metamodel

The next element we look at is the material element.

Understanding the material element

The **material** element is a passive structure element that is used to represent raw material, physical products, or energy (https://pubs.opengroup.org/architecture/archimate3-doc/chap11.html#_Toc10045437).

We use the material element to represent the products we distribute. The material icon is a hexagon, as shown here:

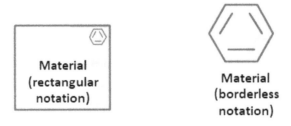

Figure 6.35 – Material icon

The material element-focused metamodel is depicted in the following diagram:

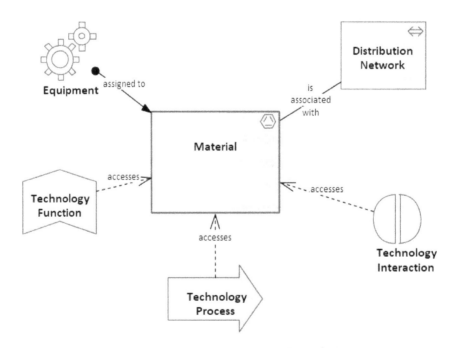

Figure 6.36 – Material-focused metamodel

The last element we will look at in this section falls into the category of generic elements.

Understanding the location element

The **location** element represents a specific physical location such as an address, street, city, state, floor, or office in a building (`https://pubs.opengroup.org/architecture/archimate3-doc/chap04.html#_Toc10045309`).

The icon for location is an inverted teardrop or map point, as shown here:

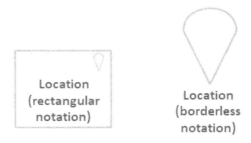

Figure 6.37 – Location icon

> **Important Note**
>
> The location element is a generic element and not specific to the physical or Technology Layer. It may be used in any ArchiMate® diagram.

The following focused metamodel shows us the relationships of the location element:

Figure 6.38 – Location-focused metamodel

The elements described thus far make up the bulk of the physical layer. Now, let's see how we can use these elements to provide some context around *ABC Trading*'s physical infrastructure.

Putting the elements together for ABC Trading

As mentioned in previous chapters, *ABC Trading* has around 1,000 employees. Their operation is distributed among four different facilities and a fleet of trucks. The following diagram provides context around those facilities:

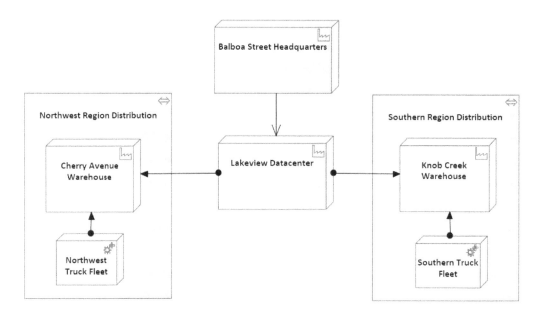

Figure 6.39 – ABC Trading facilities

The preceding diagram is an example of a context diagram. The intent is to show the major facilities at *ABC Trading* and their relationships. The diagram uses the facility, distribution network, and equipment element types. The primary link types used are the **serving** and **assignment** links.

As the diagram depicts, **Lakeview Datacenter** is assigned to each of the warehouses and serves the headquarters. Each truck fleet equipment element is part of a distribution network that serves its respective warehouse. Besides the new elements from the physical layer, there is something different about this diagram—something that we have yet to cover in this book: different ways to model dependencies.

Modeling dependencies in ArchiMate® 3.1

You may have noticed that the warehouse facility elements and the truck fleet equipment elements are nested inside of their respective network distribution elements. ArchiMate® 3.1 provides for two different means of representing aggregation and composition dependencies. You may choose to use the standard link type or a nesting relationship, as depicted in the following diagrams:

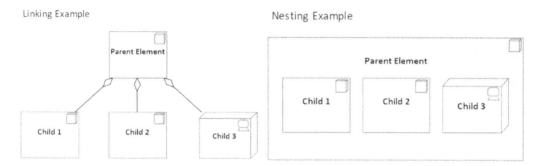

Figure 6.40 – Linking versus nesting examples

While the ArchiMate® 3.1 standard defines these two methods to mean the same thing, it's important to note that Sparx doesn't necessarily treat them the same. Establishing a link between two elements in Sparx, as in *Figure 6.40*, **Linking Example** creates a data element in the Sparx repository that holds information about the relationship that is independent of both the source and target elements being linked. In this case, Sparx does not know about the parent-child nature of the relationship.

On the other hand, in the case of **Nesting Example**, Sparx treats the **Parent Element** as a container to hold the child elements. In this scenario, however, no separate link data element is created in the repository, as we can see here:

Figure 6.41 – Nested relationship

The importance of the difference between these approaches will become clear when we get to *Chapter 11*, *Publishing Model Content*.

> **Important Note**
>
> If you are unsure about which type of relationship to use, it's probably best to stick with a link type as you will have much more flexibility in refining information about the relationship.

The following ArchiMate® 3.1 technology diagram provides us a deeper look into a distribution network:

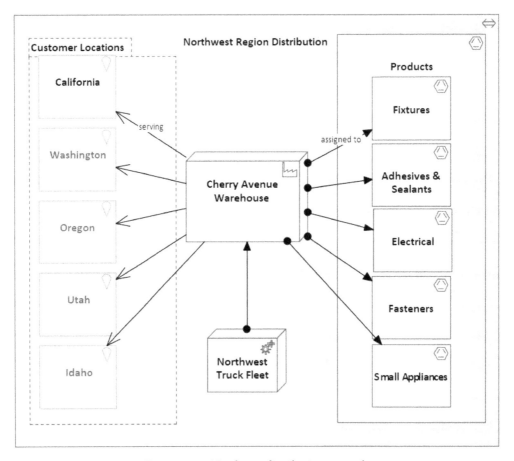

Figure 6.42 – Northwest distribution network

The preceding diagram provides a slightly deeper look into the northwest distribution network. It shows both the product types stored and customer service locations serviced by the distribution network. Note that **Cherry Avenue Warehouse**, **Northwest Truck Fleet**, and **Northwest Region Distribution** are the same elements that we created in *Figure 6.39*.

With two simple physical layer diagrams, we've provided an overview of all the relevant facilities, equipment, products, and locations involved in our business. Now, let's look at how we keep these facilities communicating through our network.

Modeling networks

In this section, we review the following element types and their relationships:

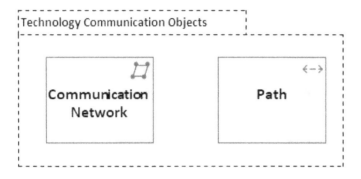

Figure 6.43 – Network-related elements

If you've been accustomed to communication network models produced in other model notations, your first impression of ArchiMate®'s coverage of the network may not be all positive. After all, there are only two elements and two link types in the ArchiMate® Technology Layer to represent a communication network. Other modeling languages include a plethora of elements to represent routers, switches, bridges, firewalls, and much more. They have links that represent specific aspects of a network, such as protocols, speeds, and connection types.

What may seem like a lack of coverage is not really a problem. The abstract nature of the technology device element type combined with the features and flexibility of Sparx provides a rich set of capabilities for network modeling in ArchiMate®. The details involved in establishing and maintaining a reliable communication network require specific skills. Just as we leave specific software design decisions to the software architect, we leave the specifics of communication network design to the network engineer. An enterprise architect rarely needs to specify much more than the capabilities of a communications network.

In this section, we will look at the ArchiMate® coverage of the communication network and how ArchiMate® can be applied to various network concerns at the enterprise level. First, let's look at what a network is.

The communication network element

According to *section 10.2.7* of the ArchiMate® 3.1 specification, "*A communication network represents a set of structures that connects nodes for transmission, routing, and reception of data.*" (`https://pubs.opengroup.org/architecture/archimate3-doc/chap10.html#_Toc10045416`)

The simplest network we can represent would connect two devices, as shown here:

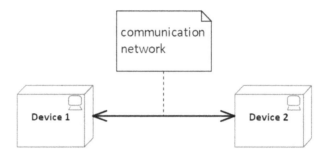

Figure 6.44 – A simple communication network

When we use the term *network* in this section, know that we are referring to a **communication network**. We need to distinguish it from other types of networks, especially a distribution network, which is part of the physical layer described in the previous section. Networks are rarely as simple as the one shown in the preceding diagram. They are far more likely to connect hundreds or thousands of devices and require many different types of hardware and software to manage.

Notice that the preceding diagram uses a connection relationship to represent the network. This connection notation makes sense because networks connect devices. In many cases, there are literally cables and wires that connect one device to another. A communication network connector is a simple solid line with arrows on either end, as in the following diagram:

Figure 6.45 – Communication network connector notation

Using a communication network connector in Sparx is a little different than making other types of model connections. Most connections in Sparx require that you mouse-drag a connector from one element to another. When you release the mouse button, you are presented with a context menu asking for the type of connection to make.

In this case, however, you need first to indicate that you are making a specific type of connection by selecting **Communication Network (Connector)** from the toolbox. Then, when you mouse-drag a connector between elements, you should not see a context menu when you release the mouse button. The communication network connector should appear automatically, as depicted in the following screenshot:

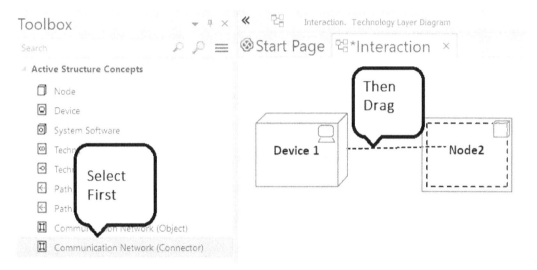

Figure 6.46 – Using a Communication Network connector

If you receive a connection context menu when you release the mouse button, something went wrong.

Of course, connecting devices through a network means that each device has a path to every other device on the network. Using such a connection notation in a diagram could become problematic, however, when we want to depict many devices or nodes connected to each other via the network. The more nodes we add to the diagram, the more cluttered and confusing it becomes, as shown here:

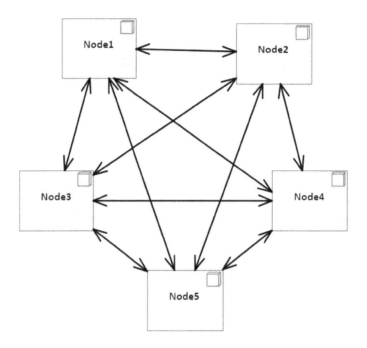

Figure 6.47 – A cluttered network diagram

To alleviate this mess, ArchiMate® 3.1 also provides an element to represent a **Communication Network**, as shown in the following diagram:

Figure 6.48 – The communication network element

Using the **Communication Network** element is very effective at reducing clutter. The following diagram is a repeat of *Figure 6.47*, but we've used the **Communication Network** element rather than links:

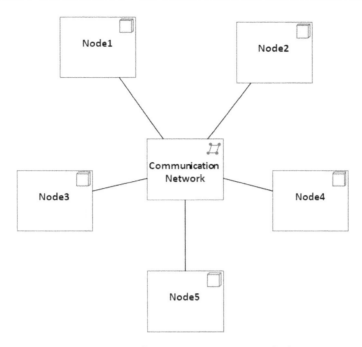

Figure 6.49 – Using the communication network element

Using element notation also provides a means of documenting aspects of the network without needing to reference specific nodes on the network. The following diagram shows some of the devices and software that make up a communication network:

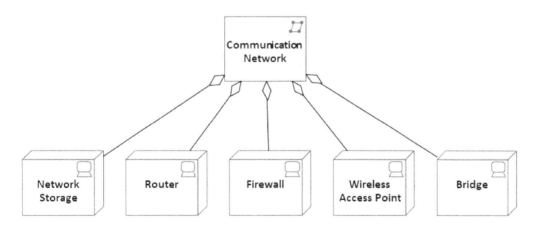

Figure 6.50 – Devices that aggregate a network

Using Sparx to model networks provides the added benefit of documenting and storing many other network attributes in the communication network element. In some cases, we use tagged values on the network element to hold various attributes of the network, as shown here:

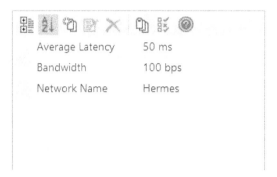

Figure 6.51 – Network attributes' tagged values

Next, we'll look at the communication network-focused metamodel.

Communication network-focused metamodel

The following diagram depicts the possible relationships of the **Communication Network** element:

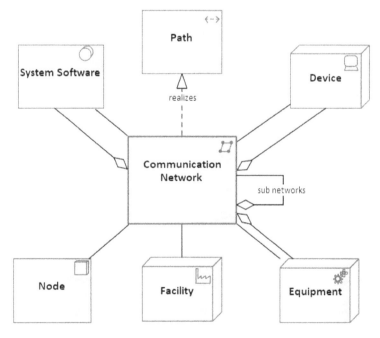

Figure 6.52 – Communication network-focused metamodel

There are three possible relationships with the **Communication Network** element. When an element needs to show that it connects to a network, we use a simple **association** relationship. If an element helps to make up a network, we use an **aggregation** relationship. The **realizes** relationship is discussed in the next section.

The path element

When your stakeholders are technical folks, especially network engineers, using a physical network view—as we have done so far—is often the best choice. *What about non-technical stakeholders?* They don't need the same level of detail, as such information is removed for the clarity of their viewpoint. Sometimes, even when our audience is technical folks, the details of a physical network view are not necessary.

> **Important Note**
>
> When details of an implementation are not necessary to communicate, it's almost always best to abstract those details away. Keep your audience focused on your point of your view.

Also, how would we represent the exchange of something other than data—say, physical items among facilities? ArchiMate® provides another way to depict a more abstract form of communication—the **Path** element and connector, as shown here:

Figure 6.53 – The path element

According to *section 10.2.6* of the ArchiMate® 3.1 specification, "*A path represents a link between two or more nodes, through which these nodes can exchange data, energy, or material.*" (https://pubs.opengroup.org/architecture/archimate3-doc/chap10.html#_Toc10045415)

The following diagram depicts two nodes connected by a path connector:

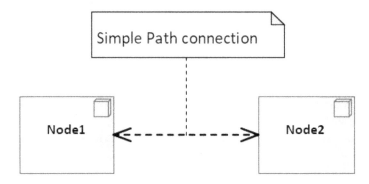

Figure 6.54 – Simple path between two nodes

The following example is more germane to our user-story exercises. In this view, the data center produces what is known as **Picking Tickets** and transmits them to each of the warehouses. A picking ticket tells a picker which items to place in a box for transport to a customer's location. In this view, the particulars of the communication network are not important:

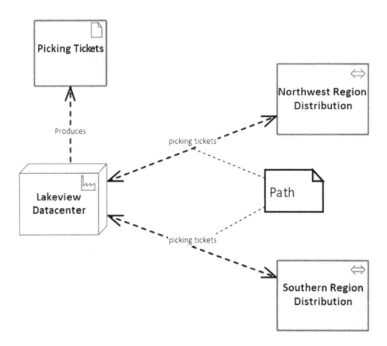

Figure 6.55 – Transfer of picking tickets to warehouses

When used as an abstraction to a communication network, the relationship between the communication network and the path is a *realization* relationship, as depicted in the following diagram:

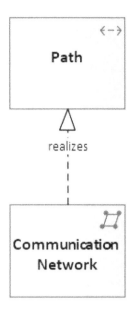

Figure 6.56 – Physical network realizes a path

A path can also be used to represent the transfer of energy, as in the following diagram:

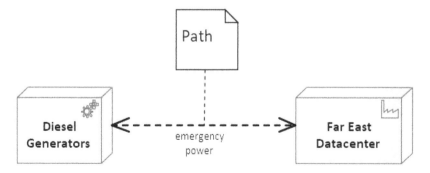

Figure 6.57 – Emergency power path

A path can be used to represent the transfer of physical elements such as products. The following diagram shows products being shipped from a warehouse to customer locations:

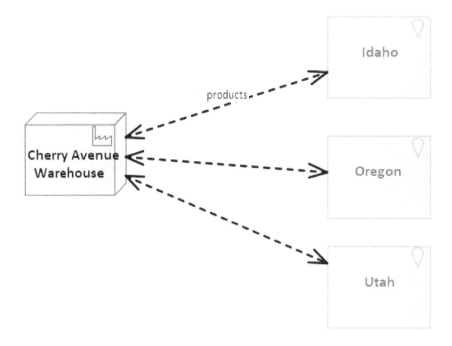

Figure 6.58 – Product shipment paths

Notice that details of how they are shipped are unimportant and thus not expressed.

Path element-focused metamodel

The logical connections that a path may make to other model elements are depicted in the following diagram:

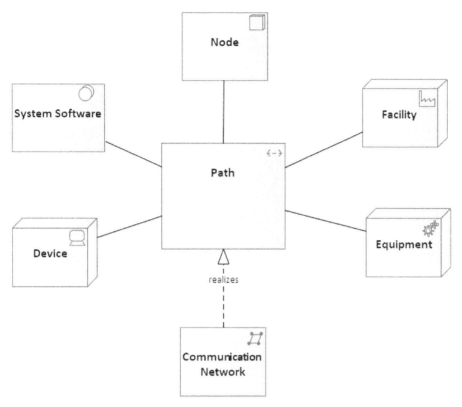

Figure 6.59 – Path-focused metamodel

As you can see, the network modeling capabilities of ArchiMate® 3.1 are quite different from other network modeling languages. Traditional networking efforts often include huge diagrams with 15 different element types and lines, traversing across other lines and off the page. They are absent from ArchiMate®. Whether such models are necessary is debatable, but certainly not at the architecture level.

Summary

In this chapter, we have covered all active and passive structural elements in the Technology Layer of ArchiMate®. That's a lot to digest in one sitting. Hopefully, we haven't overwhelmed you.

To recap, we've learned how to model technology environments, communication networks, and physical environments. We also picked up some tips on using diagram layers and tag values in Sparx. Practicing the techniques presented here provides the skills necessary to communicate the most important aspects of your technology environment without losing your audience in the details. Using the features of a tool such as Sparx can help you do this while still maintaining the details necessary for implementation.

In the next chapter, we will cover the behavioral elements of the Technology Layer. We will also use some of what we've learned here to take a deeper look at *ABC Trading's* enterprise.

7
Enterprise-Level Technology Architecture Models

As we turn our attention to *user story 3*, we see that all of its activities involve technology, and our primary audience is the **Chief Technology Officer** (**CTO**). This might lead us to believe that we need some highly technical models. But that is not true!

While the subject matter is technology, as with other models, the level of detail in the technology layer model should be tailored to the audience. Unlike previous user stories, this one requires that we have an enterprise view of technology. That means we need to examine all of the existing technologies at *ABC Trading*. We need to understand where they are used and for what purposes.

In this chapter, we will first look at the behavioral elements of the **ArchiMate® 3.1** specification and how they are used. We will then turn our attention to building the catalogs we will need to identify duplicate technologies. The sections covered in this chapter are as follows:

- Using technology behavioral elements
- *ABC Trading*'s technology background
- Building the technology components catalog
- Modeling technology services
- Reporting our findings

In this chapter, you will learn how each of the technology behavioral elements works. You will also learn how to reuse existing information by importing elements into Sparx, and how to establish relationships among elements without the need to diagram them. Finally, you will get a peek at some of the ways to present our findings. Let's get started! But first, let's talk about what you need for this chapter.

Technical requirements

In addition to having Sparx installed, you should be familiar with and have access to Microsoft Excel and **Comma-Separated Values** (**CSV**) files. It would also be helpful to have familiarity with the technologies typically employed in a data center environment, although it is not mandatory.

The process described in this chapter for importing data uses the **Sparx-native CSV Import function**. A more direct route exists for those who have purchased the **Microsoft Office extension**. If you would like to practice importing elements yourself, the **Excel** worksheets and .csv files can be downloaded from GitHub at the following links:

- https://github.com/PacktPublishing/Practical-Model-Driven-Enterprise-Architecture/blob/main/Chapter07/ABC%20Trading%20Hardware.xlsx

- https://github.com/PacktPublishing/Practical-Model-Driven-Enterprise-Architecture/blob/main/Chapter07/ABC%20Trading%20Hardware%20prepped-4-Import.csv

- https://github.com/PacktPublishing/Practical-Model-Driven-Enterprise-Architecture/blob/main/Chapter07/ABC%20Trading%20Software%20prepped-4-Import.csv

- `https://github.com/PacktPublishing/Practical-Model-Driven-Enterprise-Architecture/blob/main/Chapter07/ABC%20Trading%20Equipment%20prepped-4-Import.csv`

We will continue adding the content of this chapter into the same Enterprise Architecture (EA) repository that we used in *Chapter 6, Modeling in the Technology Layer*. If you have not read that chapter, we strongly advise you to read it first and then come back to this chapter. If you want, you can download the repository file of this chapter from GitHub instead of starting from scratch. Some of the steps in this chapter depend on elements that have already been created in the repository, so it is better to not start this chapter with an empty repository.

We will use the following ArchiMate® 3.1 specification chapters to guide our development:

- *Chapter 5, Relationships* (`https://pubs.opengroup.org/architecture/archimate3-doc/chap05.html#_Toc10045310`)

- *Chapter 10, Technology Layer* (`https://pubs.opengroup.org/architecture/archimate3-doc/chap10.html#_Toc10045407`)

- *Chapter 12, Relationships Between Core Layers* (`https://pubs.opengroup.org/architecture/archimate3-doc/chap12.html#_Toc10045440`)

Using technology behavioral elements

This section covers the **behavioral elements** within the **ArchiMate technology layer**. The following diagram illustrates the elements covered:

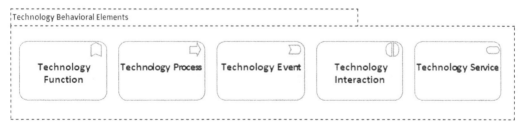

Figure 7.1 – The technology behavioral elements

If you read *Chapter 5, Advanced Application Architecture Modeling*, the technology behavioral elements will be familiar to you. In fact, the technology layer behavioral elements even use similar names. The primary difference between behavioral elements of the technology layer and those of the application layer is the scope of the elements to which they are applied. A technology layer behavioral element works only within the technology layer domain. Except for the Technology Service element, all behavioral elements in the technology layer represent behavior that is internal to technology structural elements.

However, unlike the application layer, we don't often build or modify technology layer components. Most of our technology is purchased from technology vendors. You may ask yourself why we need to describe how a technology component performs a function when that behavior is hidden from us. The short answer is that, unless your organization is in the business of developing technology to sell, you do not need to do this often.

However, there is one important exception to this fact. Sometimes, describing *how* something is done provides clarity to *what* is being done. It's important to note that behavioral elements are abstracted from the actual implementation. After all, we really don't know how most technologies are implemented, but we can describe how a function is performed in abstract terms. Let's look at an example.

At *ABC Trading*, we need to propose a new inventory management system. Our current system maintains data in several different flat-file formats. Our current application is doing a great deal of extra work just to piece together an existing inventory item. Adding new items is slow and error-prone. Our proposed new system places all data in a **Relational Database Management System** (**RDBMS**) and will be able to add an inventory item by executing a single **Structured Query Language** (**SQL**) statement. We need to describe what behavior is happening behind that single SQL statement. This becomes especially helpful for stakeholders who are familiar with how the existing system currently works.

Understanding the responsibilities of a relational database, such as key field lookups, index references, index maintenance, cluster selection, transaction management, and failover scenarios, can go a long way toward justifying our proposed project. We need to consider our audience. Our CFO must approve all such projects, but they are not a technology person. The benefits of an RDBMS may not be as obvious to them as it is to developers at *ABC Trading*. Even for technical folks, you may need to describe the specific behavior you are counting on from the single SQL statement.

The technology layer behavioral element that you will likely use most often is the Technology Service element. That's because this is the only element that exposes the internal behavior of technology components. Let's take a look at this element first.

Using the Technology Service element

According to ArchiMate®, *"A technology service represents an explicitly defined exposed technology behavior. A technology service exposes the functionality of a node to its environment. This functionality is accessed through one or more technology interfaces. It may require, use, and produce artifacts. A technology service should be meaningful from the point of view of the environment; it should provide a unit of behavior that is, in itself, useful to its users, such as application components and nodes"* (`https://pubs.opengroup.org/architecture/archimate3-doc/chap10.html#_Toc10045423`).

The following diagram serves as our focused metamodel for the **Technology Service** element:

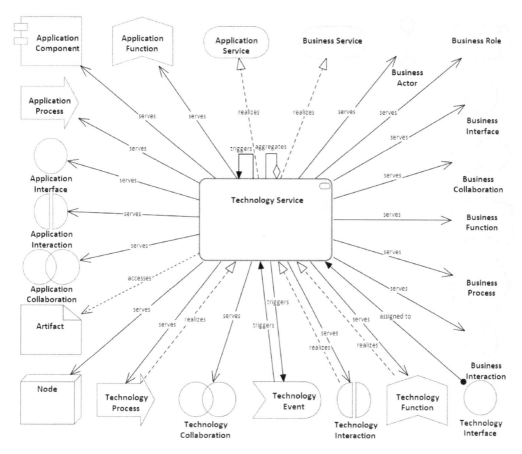

Figure 7.2 – The Technology Service-focused metamodel

As you can see, a technology service can serve just about every other element in ArchiMate®. Technology services can also help to realize an application service and a business service. Examples of technology services include reading data, storing data, connecting to a resource, sending messages, receiving messages, and much more. We will use the Technology Service element extensively in the *Modeling technology services* section of this chapter, so we will hold our examples until then.

Probably the next most common technology layer behavioral element we will use is the technology function. Let's take a look at it next.

Using the Technology Function element

According to ArchiMate®, *"[a] technology function represents a collection of technology behavior that can be performed by a node"* (`https://pubs.opengroup.org/ architecture/archimate3-doc/chap10.html#_Toc10045419`).

The following focused metamodel shows the possible relationships of the **Technology Function** element:

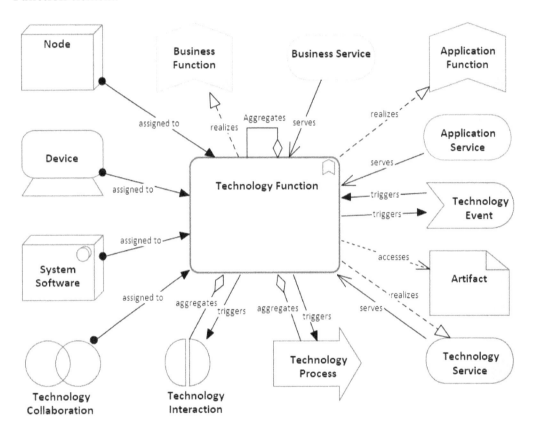

Figure 7.3 – The Technology Function-focused metamodel

The technology function is internal to a node. It can be exposed only through a technology service. For a simple example, we will use a familiar diagram, *Figure 6.5*, from the previous chapter to illustrate its use. In the following diagram, we have added Technology Function elements to describe the roles of the network firewall and the **Router** elements of the LAN:

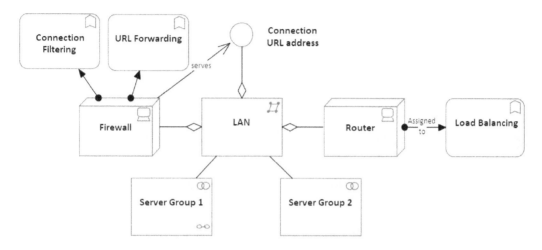

Figure 7.4 – The firewall and router behaviors

From the preceding diagram, we see that the network firewall is assigned to provide **Connection Filtering** and **URL Forwarding**. The network router provides **Load Balancing**. These functions are internal to the **Firewall** and **Router** elements. In these examples, we don't know how the firewall performs connection filtering or URL forwarding, nor in what order these functions are performed. Those are implementation details that we don't need and are not privy to.

If you've got a sharp eye, you may notice something a little odd about *Figure 7.3*. In ArchiMate®, most of the relationships that cross the level boundaries follow a distinct pattern. Upper levels tend to depend on lower levels and lower levels tend to serve upper levels, as depicted in the following diagram:

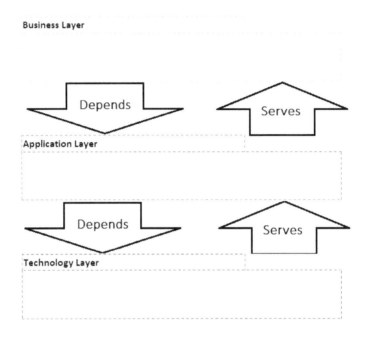

Figure 7.5 – ArchiMate® inter-layer relationships

However, *Figure 7.3* depicts a business process servicing a technology function. This means the technology function is dependent on business service. How does this make sense? Indeed, *Figure 104* of *section 12.1* of the ArchiMate® 3.1 standard implies this relationship (`https://pubs.opengroup.org/architecture/archimate3-doc/chap12.html#_Toc10045441`). But it does so by using the abstract *internal behavior element*, so its specific meaning may not be clear.

Honestly, we weren't sure ourselves, so we reached out to *The Open Group®* for clarification. The answer we received makes perfect sense. In cases where a technical node or behavior consumes the services provided by an external entity or organization, we may not be privy to, or interested in, the technical details of that service. There is no distinct element in ArchiMate® to represent an external entity; therefore, we can show the external entity's business service servicing our technical layer element. A similar relationship exists between technical elements and application services.

Next, we will look at another behavioral element – the technology process.

Using the technology process

"A technology process represents a sequence of technology behaviors that achieves a specific result" (`https://pubs.opengroup.org/architecture/archimate3-doc/chap10.html#_Toc10045420`).

The following focused metamodel shows the possible relationships of the **Technology Process** element:

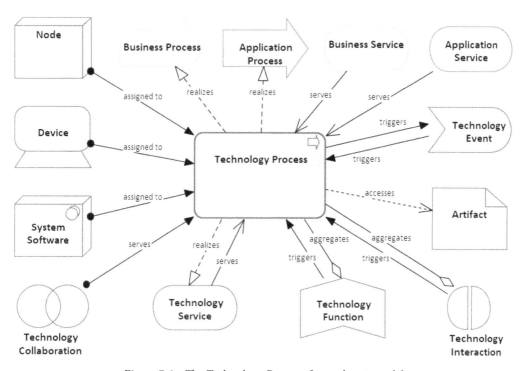

Figure 7.6 – The Technology Process-focused metamodel

The key distinction between the technology process and other behavioral elements is the notion of a *sequence* of behaviors. The technology process is a behavior that requires the execution of multiple functions or steps to be completed, often in some order. The following diagram illustrates one usage of the process element:

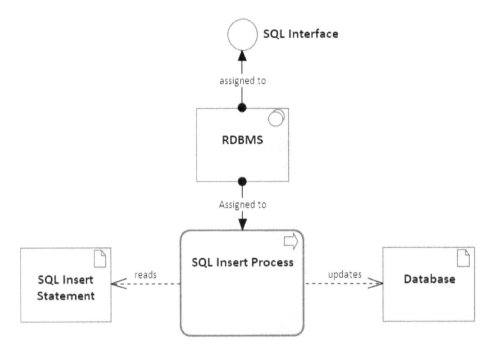

Figure 7.7 – A technical process element example

In this example, **SQL Insert Process** is performed internally by the RDBMS software node. The process reads a SQL INSERT statement and updates a database accordingly. Relational databases perform several operations when processing such statements. If it's important to document those steps, Sparx provides an easy way to do so. The following diagram shows the **Properties** dialog for the **SQL Insert Process** element:

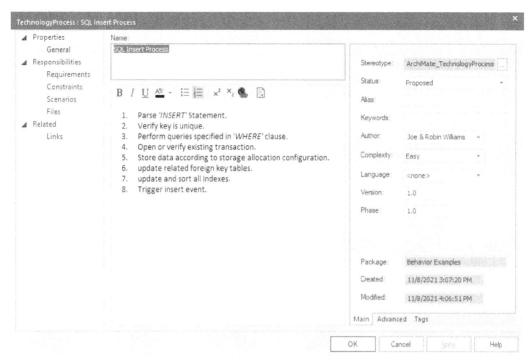

Figure 7.8 – The SQL Insert Process element Properties dialog

In the **Notes** section of the **Properties** dialog, we have described a series of actions performed by the database software when performing the insert process. The actions are described in very general terms. Any given database engine may perform these or other steps. Also, they are not necessarily performed in this precise order. Most likely, you will not be describing the actions of a process at all. Knowing that it is a process in itself is often detailed enough.

The next technology behavioral element we will look at is used to describe a more involved relationship – it's the technology interaction element.

Using the technology interaction element

"A technology interaction represents a unit of collective technology behavior performed by (a collaboration of) two or more nodes" (https://pubs.opengroup.org/ architecture/archimate3-doc/chap10.html#_Toc10045421).

A **technology interaction** occurs between two or more node elements to achieve a specific function or behavior, as shown here:

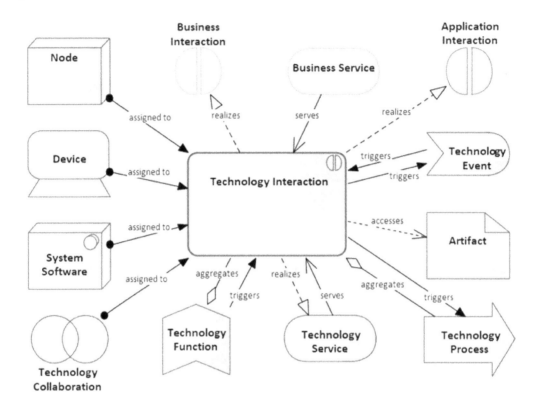

Figure 7.9 – The Technology Interaction-focused metamodel

Technology Interaction elements are logically paired with Technology Collaboration elements. In this sense, the technology collaboration identifies the nodes that participate in the interaction, whereas the technology interaction describes the function that the collective nodes are to perform. For example, a database engine node and an application server node may collaborate to manage database update transactions, as shown in the following figure:

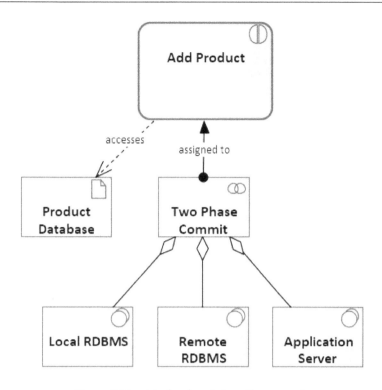

Figure 7.10 – A technology interaction example

In the preceding example, the **Add Product** interaction accesses the product database for an update and performs a two-phase commit, which includes a local database engine, a remote database engine, and an application server. The **Add Product** interaction may also serve to realize the add product service (not depicted).

The final technology behavioral element we will look at is the Technology Event element.

Using the Technology Event element

A **technology event** represents a change of state within the technology environment. In short, something has happened in the environment that is important for some reason. Behavioral elements may trigger an event or may be invoked by an event. Structural elements such as nodes may be assigned to an event, implying that they will respond to the event in some way. In event processing systems, we say the nodes may *subscribe* to an event.

"Technology functions and other technology behavior may be triggered or interrupted by a technology event. Also, technology functions may raise events that trigger other infrastructure behavior. Unlike processes, functions, and interactions, an event is instantaneous: it does not have duration. Events may originate from the environment of the organization, but also internal events may occur generated by, for example, other devices within the organization" (`https://pubs.opengroup.org/architecture/archimate3-doc/chap10.html#_Toc10045422`).

Technology events can have the relationships depicted in the following focused metamodel:

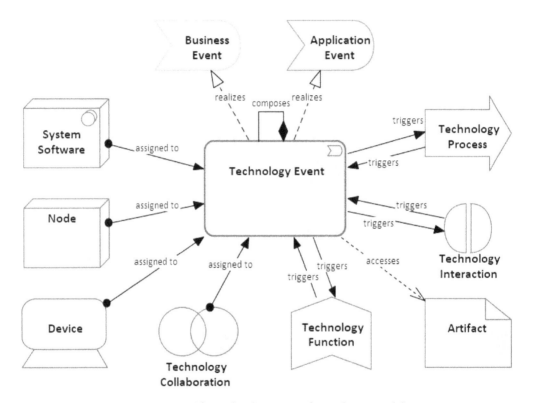

Figure 7.11 – The Technology Event-focused metamodel

Examples of events include opening or closing a file, connecting to a resource, starting or completing a process, starting or stopping a device, or updating a database table.

In the following example, when a new product is added, it triggers a **Propagate Changes** interaction to propagate the change across all database instances:

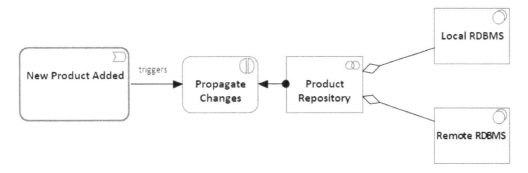

Figure 7.12 – A technology event example

The elements discussed in this section represent the core of the technology behavioral elements. Before we move on to building our technology catalogs, let's take a look at why we're doing this in the first place and gain some context around *ABC Trading*'s technology background.

ABC Trading's technology background

In our scenario, the CTO is new to the organization. They are under pressure to control technology costs. Like many organizations, *ABC Trading* has dozens of applications that have been developed or purchased at various points in its history. Along with each application, new technology components have been implemented to support those applications and various other projects.

The CTO suspects that, along the way, some applications and technology components that are no longer needed have been left in place. It's also possible that the use of some applications or components can be eliminated by making relatively small changes. Unlike the previous user stories, this one requires us to analyze information across a broad spectrum of the enterprise rather than looking at the details of a single application or technology. The problem is that the CTO does not have this information.

But why not? It's not as if systems are implemented haphazardly. *ABC Trading* has a robust change management process in place. This process requires that new applications come with detailed design information in the form of Microsoft Word documents. Changes to the data center and network environment go through the same process. The information is generated at implementation time. Since this process has been in place for several years, you would think that all of the necessary information is available.

The simple answer is that the information does, indeed, exist. The applications development team keeps scores of design documents from years of changes. Database administrators maintain data models in conceptual, logical, and physical forms. The operations department maintains a spreadsheet of all servers and physical equipment in place in the data centers. The network folks are constantly updating their network infrastructure diagram. So, what's the problem?

One problem is that this information is rarely reused. When the application team designs a change to an application, they make their own reference to the database or network environments, rather than using the information generated by the operations or network folks. It's just easier that way.

When the operations team renames or replaces a server, they never go back and change views that rely on them. Each department maintains its independent perspective of the enterprise. Rarely are any of the views completely accurate. From the enterprise perspective, this information is unreliable. It's simply not trusted. This is especially frustrating, given the amount of time and work devoted to creating and maintaining documentation. It's our job as **Enterprise Architects** to integrate this information into an enterprise repository from an enterprise perspective. We need enterprise-level information. To keep it accurate and reliable, we need to build trust in our enterprise information.

Confidence in enterprise information will not be realized quickly but will come gradually. In *Chapter 10*, *Operating the EA Repository*, we will look at ways to ensure the integration of information across the enterprise. The following section describes one convenient way to establish the core of the technology layer, the **technology components catalog**.

Building the technology components catalog

To address the CTO's concerns about redundant and unused technology, we need information in our repository for all equipment at *ABC Trading*. From the repository, we will identify the services provided by each technology. Where two or more technologies provide the same service, we have a potential duplicate. We can then analyze the usage of each technology to determine whether we might be able to eliminate one of them. First, let's look at what the technology components catalog is.

Defining the technology components catalog

Simply put, the technology components catalog is a list of all hardware, system software, and equipment in use at *ABC Trading*. Examples of hardware include servers, storage devices, networking devices, and so on. Examples of equipment include server racks, coolers, humidifiers, sensors, and much more.

System software includes such things as operating systems, database software, storage or backup systems, application containers, transaction monitors, and security software. For our purposes, the technology components catalog must reside in our repository in Sparx. This is where we will be performing our analysis and reporting our results.

Collecting the information

One possible way to build this catalog in Sparx is to fully document each application in the enterprise, including all technology components used by the application. Doing this will give us a complete list of each technology component in use. While it would be wonderful to have such information in Sparx, there are at least two problems with this approach:

- Firstly, reporting from an application perspective will not reveal technologies not being used by any application.
- Secondly, documenting each application would take an enormous amount of time and effort.

Certainly, the CTO would like this information before they retire. So, we will need another, more practical, approach.

It is common for organizations to maintain information in spreadsheets. We need to reuse this information where practical. Fortunately for us, the data center manager maintains an inventory of equipment housed in the data center. Each of the warehouse managers has a small list of equipment in use at their respective locations. After meeting with these folks, we have most of the information we need, but it's in the form of a set of Microsoft Excel spreadsheets.

Spreadsheets are great for crunching numbers, and knowledge of Microsoft Excel is ubiquitous in the industry. However, when knowledge of complex relationships needs to be communicated, spreadsheets often fall short of the task. This is where a modeling tool such as Sparx Systems Enterprise Architect shines.

So, we need to do some formatting to the sheets and save them in the `.csv` format. We will then build an import specification in Sparx and, finally, import the files into Sparx.

By far, the data center equipment list is the biggest. The warehouse equipment list was small enough that we simply added the warehouse equipment to the data center list. The first worksheet contains a list of the server hardware used at the data center, which looks like the following screenshot:

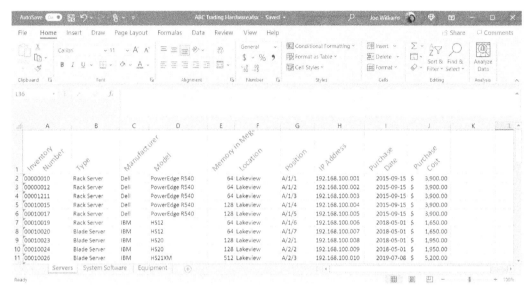

Figure 7.13 – The original hardware inventory Excel spreadsheet

> **Important Note**
>
> If you're following along with this exercise, you'll need to create a backup of the worksheet before you start.

We need to modify this list to prepare it to be imported into Sparx. Columns in the spreadsheet need to align with fields in Sparx. The following subsections detail this process.

Formatting the CSV file

First, we need to split any merged cells before attempting the import process because merged cells can confuse Sparx. The element fields we need to create are as follows:

- Name
- Notes
- Type
- Stereotype

- Tag-value field – `Location`
- Tag-value field – `PurchaseDate`
- Tag-value field – `PurchaseCost`

Tag value fields are user-defined fields that can be added to any element type without being constrained by the field data type. You can add as many tag-value fields as needed, and you need to provide a name and a value pair for each.

Formatting the element name

The first model element field value we need to create is the **element name**. The worksheet fields we will use to create this value include the following:

- `Manufacturer`
- `Type`
- `Inventory number`

Using the inventory number will make the element name unique. Sparx does not require the `name` field to have a unique value, but for readability, we would like the name to be unique – for example, the first server on the list will have the name `Dell Rack Server 00000010`. To create this new `name` column, follow these steps:

1. Insert a new column to the left of column **A**. Give the new column **A** the heading `Name`.

2. In the second row of the new **Name** column, enter the `=$C2&" "&$B2&" "&$A2` formula. This instructs the spreadsheet to set the value of the current cell to the value in cell `C2` and then append a space and the contents of cell `B2`, followed by a space and the content of cell `A2`. The result should be the name **Dell Rack Server 00000010** displayed in row **2**, column **A**.

3. Now, simply copy this new formula and paste it into all the remaining rows of column **A**. This column should now show the names of each of the servers in the data center.

4. While Excel now displays the name, the actual value of each cell is still a formula. Replace the formula in each cell with the actual text resulting from the formula. Copy the entire column **A** and paste it back in the same location using the **paste-as-text** option.

This is a recommended way to construct a meaningful name that is both readable and understandable by any user. You may decide to follow a different naming approach according to your environment and stakeholders' requirements. The next element field we need to create is the `Notes` field.

Formatting the Notes field

As you have seen in previous chapters, the `Notes` field is used to add descriptive text to any element type. The steps for formatting the `Notes` field are similar to the steps used for formatting the `Name` field. You need to replace the formula with one that better fits your needs:

1. Create a new column to the right of column **A**. Give it the name `Notes`.

2. In the second row, enter the formula to concatenate the **Model**, **Memory**, **Position**, and **IP Address** fields. We include labels for the position and IP address – `=$F2&" "$G2&"mb Position"&$H2&" IP address: "&$I2`.

3. Copy the formula and paste it into the remaining rows.

4. Convert the formulae to text by repeating *step 3* of the previous section.

5. Remove the original columns by repeating *step 4* of the previous section.

You are free to add any text in the `Notes` field, but you need to make sure that your notes are sufficiently descriptive yet not too long. We suggest limiting your notes to fewer than three lines, but it is totally up to you. Next, we need the element `Type` field.

Identifying the element types

In Sparx, every element is identified by a type and an optional stereotype. The way that Sparx works is by extending the basic UML types, such as **class** and **component**, with stereotypes. If you are modeling with ArchiMate® notations, as we do in this book, Sparx requires you to provide the stereotypes as well. Providing these two fields is required by Sparx for importing spreadsheets as ArchiMate® elements. In the sample spreadsheet that we are using, all the elements have a basic element `class` type and the `ArchiMate_Device` stereotype. We add these two columns as follows:

1. Create a new column to the right of the **notes** column. Give it the name `Type` and enter the `class` value in all rows.

2. In the same manner as the **type** column, add the **stereotype** column. Insert another column to the right of the **type** column and add the `ArchiMate_Device` value to each row.

This tells Sparx the type and the stereotype to assign for each element that will be imported. The last step is to format the `tag` fields.

Formatting the tag fields

Leave the `Location` and `PurchaseDate` fields as they are, but format the `PurchaseCost` field as `Number` with no commas to mark the *thousands* position. The import specification will indicate that these fields go into tag-value fields. The final spreadsheet should look like this:

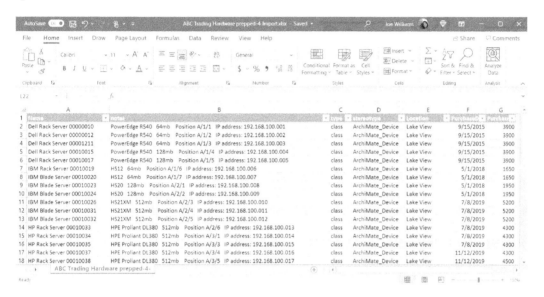

Figure 7.14 – The reformatted spreadsheet

Finally, save the spreadsheet in the `.csv` format – **File** > **Export** > **Change File Type** > **CSV (Comma Delimited)**. You may get a warning about losing content when saving to this format; that's okay. Give it the name `ABC Trading Hardware prepped-4-Import.csv`.

Caution

None of the fields can contain embedded commas. This is important, as it will cause the import to fail.

The `.csv` files are plain text files. You can view them in any text editor. Our prepped file now looks like the following:

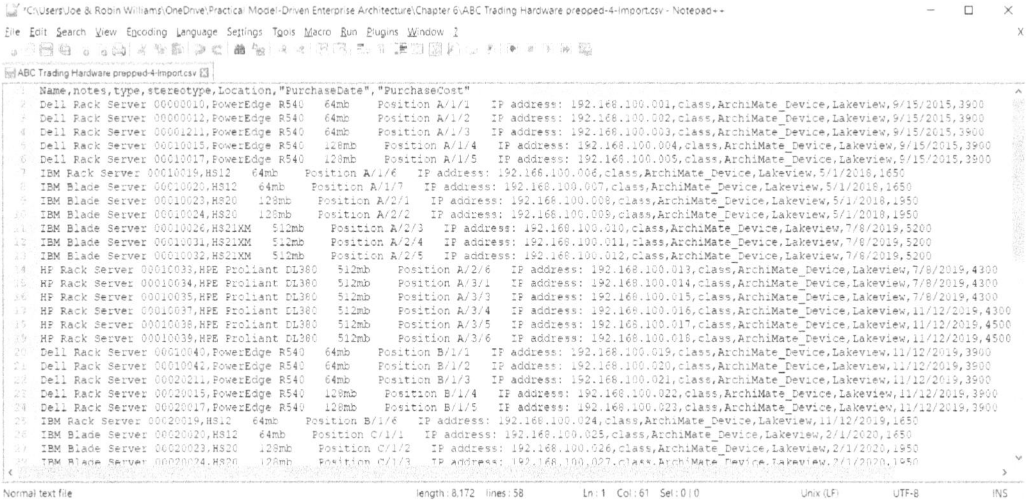


```
Name,notes,type,stereotype,Location,"PurchaseDate","PurchaseCost"
Dell Rack Server 00000010,PowerEdge R540   64mb    Position A/1/1   IP address: 192.168.100.001,class,ArchiMate_Device,Lakeview,9/15/2015,3900
Dell Rack Server 00000012,PowerEdge R540   64mb    Position A/1/2   IP address: 192.168.100.002,class,ArchiMate_Device,Lakeview,9/15/2015,3900
Dell Rack Server 00001211,PowerEdge R540   64mb    Position A/1/3   IP address: 192.168.100.003,class,ArchiMate_Device,Lakeview,9/15/2015,3900
Dell Rack Server 00010015,PowerEdge R540   128mb    Position A/1/4   IP address: 192.168.100.004,class,ArchiMate_Device,Lakeview,9/15/2015,3900
Dell Rack Server 00010017,PowerEdge R540   128mb    Position A/1/5   IP address: 192.168.100.005,class,ArchiMate_Device,Lakeview,9/15/2015,3900
IBM Rack Server 00010019,HS12   64mb    Position A/1/6   IP address: 192.168.100.006,class,ArchiMate_Device,Lakeview,5/1/2018,1650
IBM Blade Server 00010020,HS12   64mb    Position A/1/7   IP address: 192.168.100.007,class,ArchiMate_Device,Lakeview,5/1/2018,1650
IBM Blade Server 00010023,HS20   128mb    Position A/2/1   IP address: 192.168.100.008,class,ArchiMate_Device,Lakeview,5/1/2018,1950
IBM Blade Server 00010024,HS20   128mb    Position A/2/2   IP address: 192.168.100.009,class,ArchiMate_Device,Lakeview,5/1/2018,1950
IBM Blade Server 00010026,HS21XM   512mb    Position A/2/3   IP address: 192.168.100.010,class,ArchiMate_Device,Lakeview,7/8/2019,5200
IBM Blade Server 00010031,HS21XM   512mb    Position A/2/4   IP address: 192.168.100.011,class,ArchiMate_Device,Lakeview,7/8/2019,5200
IBM Blade Server 00010032,HS21XM   512mb    Position A/2/5   IP address: 192.168.100.012,class,ArchiMate_Device,Lakeview,7/8/2019,5200
HP Rack Server 00010033,HPE Proliant DL380   512mb    Position A/2/6   IP address: 192.168.100.013,class,ArchiMate_Device,Lakeview,7/8/2019,4300
HP Rack Server 00010034,HPE Proliant DL380   512mb    Position A/3/1   IP address: 192.168.100.014,class,ArchiMate_Device,Lakeview,7/8/2019,4300
HP Rack Server 00010035,HPE Proliant DL380   512mb    Position A/3/3   IP address: 192.168.100.015,class,ArchiMate_Device,Lakeview,7/8/2019,4300
HP Rack Server 00010037,HPE Proliant DL380   512mb    Position A/3/4   IP address: 192.168.100.016,class,ArchiMate_Device,Lakeview,11/12/2019,4300
HP Rack Server 00010038,HPE Proliant DL380   512mb    Position A/3/5   IP address: 192.168.100.017,class,ArchiMate_Device,Lakeview,11/12/2019,4500
HP Rack Server 00010039,HPE Proliant DL380   512mb    Position A/3/6   IP address: 192.168.100.018,class,ArchiMate_Device,Lakeview,11/12/2019,4500
Dell Rack Server 00010040,PowerEdge R540   64mb    Position B/1/1   IP address: 192.168.100.019,class,ArchiMate_Device,Lakeview,11/12/2019,3900
Dell Rack Server 00010042,PowerEdge R540   64mb    Position B/1/2   IP address: 192.168.100.020,class,ArchiMate_Device,Lakeview,11/12/2019,3900
Dell Rack Server 00020211,PowerEdge R540   64mb    Position B/1/3   IP address: 192.168.100.021,class,ArchiMate_Device,Lakeview,11/12/2019,3900
Dell Rack Server 00020015,PowerEdge R540   128mb    Position B/1/4   IP address: 192.168.100.022,class,ArchiMate_Device,Lakeview,11/12/2019,3900
Dell Rack Server 00020017,PowerEdge R540   128mb    Position B/1/5   IP address: 192.168.100.023,class,ArchiMate_Device,Lakeview,11/12/2019,3900
IBM Rack Server 00020019,HS12   64mb    Position B/1/6   IP address: 192.168.100.024,class,ArchiMate_Device,Lakeview,11/12/2019,1650
IBM Blade Server 00020020,HS12   64mb    Position C/1/1   IP address: 192.168.100.025,class,ArchiMate_Device,Lakeview,2/1/2020,1650
IBM Blade Server 00020023,HS20   128mb    Position C/1/2   IP address: 192.168.100.026,class,ArchiMate_Device,Lakeview,2/1/2020,1950
IBM Blade Server 00020024,HS20   128mb    Position C/1/3   IP address: 192.168.100.027,class,ArchiMate_Device,Lakeview,2/1/2020,1950
```

Figure 7.15 – The CSV file contents

The file is ready to be imported into Sparx, so let's see how to do it.

Importing the CSV file into Sparx

Now the fun begins! We are going to import the `.csv` file we just created into Sparx. This process consists of three steps:

1. Creating a package in the repository to hold new elements
2. Creating an import specification for the new `.csv` file
3. Executing the import using the new specification

Let's start by creating the package. You should be already familiar with creating packages from *Chapter 3*, *Kick-Starting Your Enterprise Architecture Repository*.

Creating the package for the enterprise view

In our enterprise repository, we already have a package for the *metamodels*, and we have other packages for specific applications or specific elements of the enterprise. However, the content that we are about to import now does not belong to a specific element, and we can consider it global or enterprise-level content. Therefore, we will create a separate package to contain all the enterprise content, including the spreadsheet content that we are about to import, by following these steps:

1. Start Sparx and open your repository.

> **Important Note**
>
> If you are using the EA repository from GitHub's Chapter 7 repository, the following steps have already been completed.

2. In the **Project** browser, right-click on the model's root node, **Architecture Content**.

3. Add a new view and name it Enterprise View.

4. Right-click on **Enterprise View** and add a new folder called Servers:

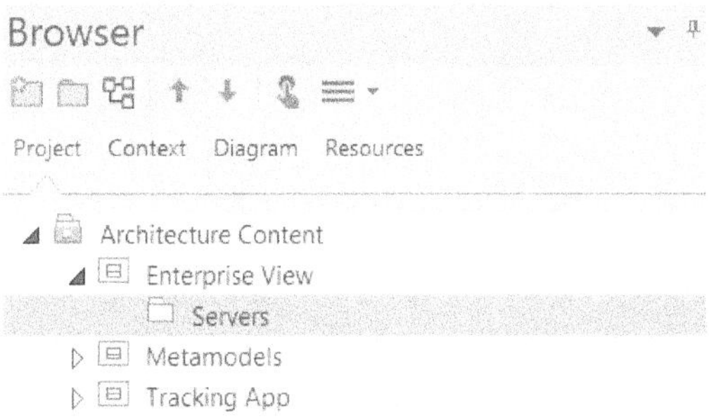

Figure 7.16 – The Servers package inside the Enterprise View package

That's good enough for now. That will hold the contents of the .csv file.

Creating an import/export specification

This process creates a specification that tells Sparx what to expect while importing our data. Go to **Publish** > **CSV** > **CSV Exchange Specification**. A dialog box will appear, asking for more information to complete the import. Then, enter the following information in the dialog box shown in *Figure 7.17*:

1. Provide a name for the specification, such as Servers.csv.

2. Select the comma from the **Delimiter** dropdown.

3. The **Notes** are optional and free-form.

4. **Default Filename** is the .csv file created earlier. When you specify this value, you will receive a warning that the file already exists. Reply **Yes** to this dialog. Don't worry – nothing will be over-written! The warning is for exporting only; we select **Import** as **Default Direction**.

5. For each of the fields that we created in the .csv file, select the field name from the **Available Element Field** section and press the **Add Field** button. The fields must be in the order that they appear in the .csv file. The following screenshot shows the **CSV Import/Export Specification** dialog and the process of adding fields:

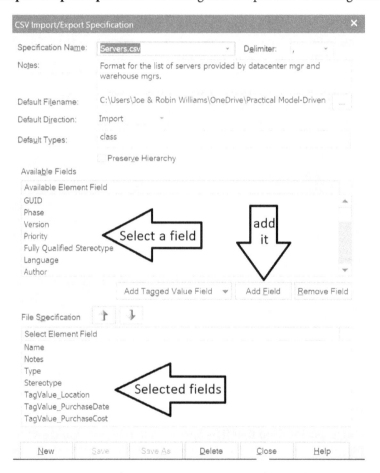

Figure 7.17 – The CSV Import/Export Specification dialog

6. For the last three fields, **Location**, **PurchaseDate**, and **PurchaseCost**, press the **Add Tagged Value Field** button. Then, select **Value**. Enter the field name and press **OK**, as shown in the following screenshot:

Figure 7.18 – Add Tagged Value Field

7. When all fields have been specified, the **CSV Import/Export Specification** dialog should look like the following:

Figure 7.19 – The completed CSV Import/Export Specification dialog

8. Press the **Save** button to save the specification data.

Now that we have created the specification, we can import our data.

Running the import function

This is where the rubber meets the road. Hopefully, all our work will pay off. This is also the easiest step:

1. In the Project browser, select the new package, **Servers**.

2. Go to **Publish > CSV > CSV Import/Export**.

3. From the dialog that displays, enter the name of the specification we just created, `Servers.csv`.

4. Make sure that the **File** field points to the `.csv` file we just created.

5. The remaining fields have default values, which we will accept.

6. Press the **Run** button. Sparx executes the `import` function and reports the status of each row processed from the import `.csv` file, as shown in the following figure:

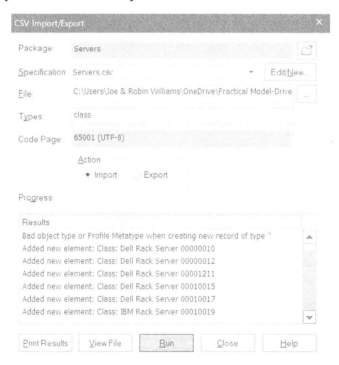

Figure 7.20 – Successful import results

7. Note that the first record reported an error, **Bad object type or Profile Metatype when creating a record of type**. This is okay. The first record is the header row. Since there is no element type called `type`, an error is reported for that record. It is a small price to pay for keeping our field names straight, but if you do not like error messages, you are free to remove the header row in future imports.

8. Click the **Close** button to close the dialog.

Next, we will see how the imported elements look in Sparx and how to fine-tune them.

Reviewing the results

Now, let's look at what we have added to the repository. Expand the **Servers** package in the Project browser. You should see a list of elements, as shown here:

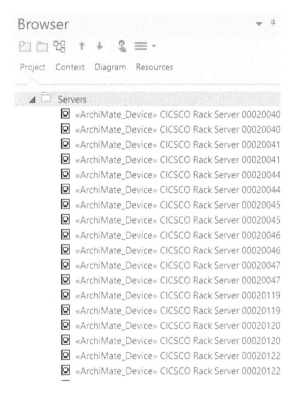

Figure 7.21 – The ArchiMate® devices

Select one of the elements in the list. Right-click and select **Properties** from the context menu, and then select **Properties** again from the context sub-menu. The **Properties** dialog will appear. Select the **Tags** tab on the right side of the dialog. The results should look something like the following:

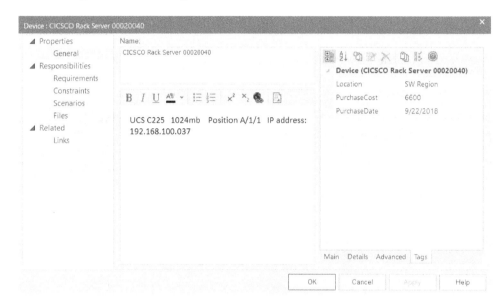

Figure 7.22 – The device properties

Now that we have imported information regarding all the servers at *ABC Trading*, we need to follow the same steps with other model elements so that we can build the information around complex relationships and, later, report on it.

Equipment and system software

This section describes other technology component elements that we need to import. We will not go into detail about how to format and input this data, as the process is the same as what has already been described. Given the equipment list, we have created another .csv file import specification called Equipment.csv and loaded that information into another repository package called Equipment. We have done the same with System Software. The Sparx Project browser now looks like this:

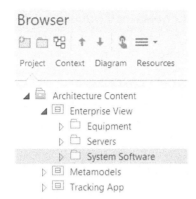

Figure 7.23 – The tree Enterprise View catalogs

If you open the **Properties** dialog for any element in the **Equipment** or **System Software** catalogs, the Properties dialog will show information like this:

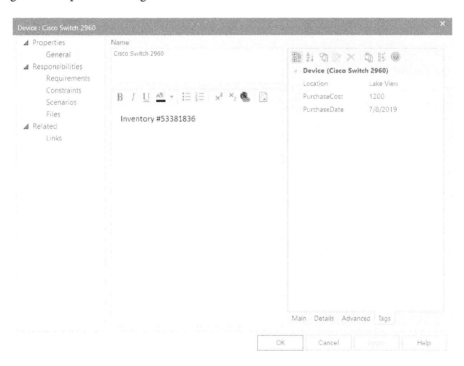

Figure 7.24 – Equipment properties

Hopefully, we've demonstrated the power of the import/export functionality within Sparx effectively. For practical reasons, we have kept these contrived examples rather small. Many organizations have many more technology assets than what we have demonstrated here. Most organizations maintain some sort of information about their technology assets. Entering these components manually would be both time-consuming and fraught with potential errors. With relatively little effort, we were able to reuse the existing information in Sparx. Now, let's see how we can use this information.

Modeling technology services

In this section, we will be exploring ways to identify potential duplicate resources within *ABC Trading*'s technology infrastructure. To do this, we will be referencing the devices and systems software elements that we imported in the last section. The process that we will use involves identifying and modeling the discrete services supplied by each of the technology elements in place.

Identifying existing services

Technology Service elements are used to describe functionality performed by the node that is made available to other elements in the environment. If you're new to the concept of technology-providing services, this process can seem a bit daunting. What are the services? What should they be called? What do they mean? Fortunately for us, the TOGAF® **Technical Reference Model** (**TRM**), `https://pubs.opengroup.org/architecture/togaf8-doc/arch/chap19.html`, provides an excellent starting point in the form of a taxonomy of technical services. To help you use that taxonomy, we have included over 100 service categories as Technology Service elements in our Sparx repository, located at `https://github.com/PacktPublishing/Practical-Model-Driven-Enterprise-Architecture/blob/main/Chapter07/EA%20Repository.eapx`. The location of these services in the repository is depicted in the following screenshot:

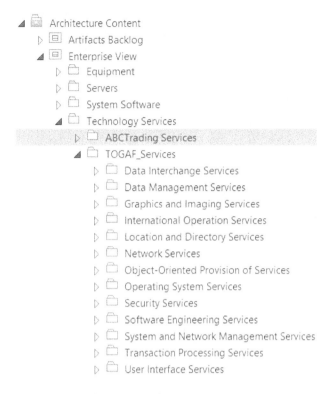

Figure 7.25 – The location of the TOGAF® service categories

While these TOGAF® technology service categories are an excellent starting point for forming your technology services catalog, they aren't ideal for demonstrating our next step; there are just too many of them. For that reason, we will use a much smaller set of services that we've created from scratch.

> **Important Note**
>
> If you are following along using the EA repository from GitHub's Chapter 7 repository, the folders have already been created in that repository.

The following is the process we will use to create our much smaller service catalog:

1. Select the **Enterprise View** element in the Project browser.

2. Click the new folder icon.

3. Name the folder Technology Services.

4. Create a sub-folder called ABC Trading Services.

5.   Click on the **New Diagram** button.

6.   Select **ArchiMate 3.1** and the **ArchiMate 3::Technology** diagram type from the **New Diagram** dialog:

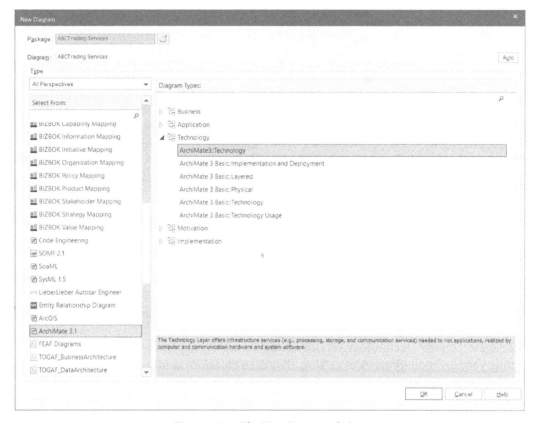

Figure 7.26 – The New Diagram dialog

7.   If you're creating your organization's technology services catalog, select the **Technology Service** element type from the toolbox and drag it onto the diagram. Name the new technology service appropriately. Perform this step for each of the services in your enterprise. If you're following along with our example, skip this step and go to the next one.

8.   If you're following along with our example, all of our technology services have already been defined in the ABCTrading Services folder. From the browser, simply select all of the technology services and drag them onto the diagram in one step.

The following diagram identifies a list of services supplied by *ABC Trading* technologies. As mentioned previously, this is an abbreviated set for demonstration:

Figure 7.27 – A technology services catalog example

> **Important Note**
>
> Creating a diagram such as the preceding one can be fast and easy. Simply select all the elements you want and drag them over. Then, select **Layout > Diagram Layout > Apply Default Layout**. You can easily try different layout types until you find the one you like.

As with every other enterprise element, technology services do not exist by themselves and must be related to other elements. In the next section, we will see how technology services are related to technology components and how we can generate useful models using this relationship.

Mapping services to technology components

We need to identify which services are provided by each node in our repository. We say that a node is *assigned* to a service, so we do this by establishing an assigned link between the node and the function. The following figure shows us how a technology node mapped to a service might look:

Figure 7.28 – Using the assignment relationship

The most obvious way for us to establish a link like this is to create a diagram, drag the elements onto the diagram, and then drag a link from one element to the next. The following figure shows an example of how we do this for device nodes:

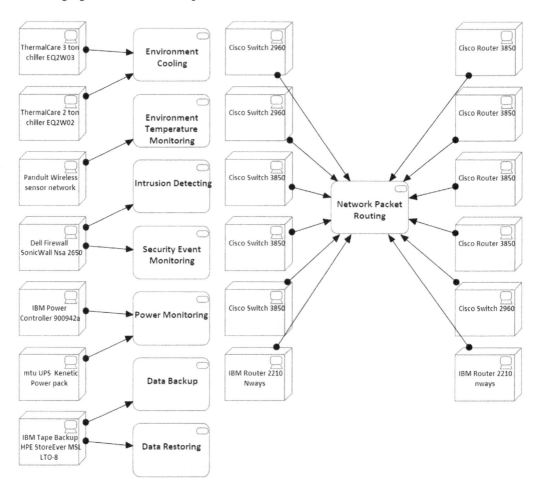

Figure 7.29 – Assigning services to devices

As you can see, the problem with this approach is that there are a lot of connections to be made. Although this is a temporary work diagram, it still becomes cluttered rather quickly. With hundreds or even thousands of technology nodes and potentially hundreds of services, creating such a diagram becomes impractical. Nor would we want to create a diagram for each element. We'll take a look at another feature of Sparx that gives us an alternative to the diagram method – the matrix feature.

Using the matrix feature

If you decide that establishing links in a diagram as described in the previous section is the way to go, you can skip this section and go to the *Reporting our findings* section. This simply explains an alternative to dragging all elements onto a diagram. This feature lets you establish links between elements using a two-dimensional matrix, with one package of elements as the rows and another package as the columns. *Figure 7.30* illustrates the matrix. The following are the steps to opening the **Relationship Matrix** tab:

1. In the Project browser, select the `System Software` folder.
2. Go to **Design** > **Matrix** > **Open as source**. The **Matrix** tab is displayed.

> **Important Note**
>
> The following screenshot of the Relationship Matrix, *Figure 7.30*, has been cropped in order to show the details. You may want to toggle full-screen mode by selecting **Start** > **Full Screen**.

3. The **Source** field should be prepopulated with the **System Software** folder name. In the **Target** field, browse and select the **ABC Trading Services** folder.
4. In the **Source Type** field, select **SystemSoftware** from the drop-down list.
5. In the **Target Type** field, select **TechnologyService**.
6. In the **Link Type** field, select **Assignment**.
7. The **Direction** field is set to **Source -> Target**:

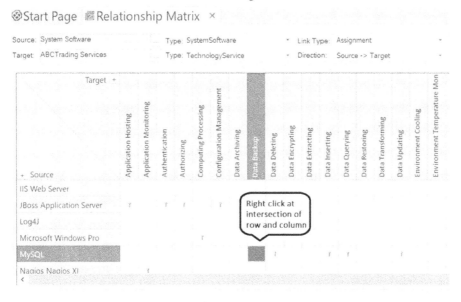

Figure 7.30 – The Relationship Matrix tab

8. Select a cell from within the matrix.

9. Right-click on the cell and click **Create new relationship** >
    ArchiMate3.1:Assignment.

> **Important Note**
>
> You may need to reposition the boundaries between the rows and columns to
> see the full element name. Simply drag the boundary to reposition it.

Once we have the matrix the way we like it, we'll save the settings in a new matrix
profile. This will allow us to close the tab and return later without reentering the
configuration information. We will be able to simply select the saved profile from
the **Profiles** drop-down list.

10. Press the **Options** button at the upper-right corner of the **Relationship Matrix** tab.

11. Select **Profiles** > **Save as New Profile**.

12. Give the new profile the name SystemSoftwareServices:

Figure 7.31 – Save the matrix configuration as a new profile

As you go through the process of creating links between nodes and services, you may
realize that a particular service element doesn't quite fulfill your expectations. You have
complete control over the model. You can rename a service or create a new one that is
slightly different to fit the need. If you do this, you'll want to define these differences in
the notes section of the model element. In fact, it's not a bad idea to document all of the
technology services this way. It may help when trying to decide whether a particular node
actually provides the service in question. Always keep the goal of the exercise in mind – in
this case, to catch redundant or unused technology.

Once we have mapped each system software node to at least one service, we will need to repeat this process for the **Equipment** and **Servers** packages. As you progress through the process of creating these links, the matrix can become quite complicated. There are options available within the following **Matrix Options** dialog that allow you to highlight rows and columns without relationships. You can access this dialog from the **Options** button in **Relationship Matrix** (see the preceding *Figure 7.31*).

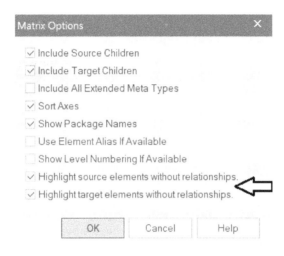

Figure 7.32 – The Matrix Options dialog

These options result in a convenient means of keeping track of your progress. If you look at the Sparx Relationship Matrix on your **User Interface** (**UI**), you will find that system software rows with no relationships to services are highlighted in blue. Services with no related system software elements are highlighted in pink.

Once we're satisfied that our matrix is complete, we will turn our attention to how to present the information.

Reporting our findings

Once again, how we present our findings depends on the audience. There are many options available. We need to try to anticipate whether our audience would be more receptive to diagrams or lists. How much information is necessary? We also need to take into consideration what will be done with this information. What are the next steps?

In our case, the CTO is rather new to the company. They are quite experienced, but they came from a somewhat different industry. Perhaps we need to add more descriptions of the technologies in question. We need to ask ourselves the following questions:

- How formal do we need to be in presenting the information?
- Do we need to introduce this subject?
- Do we need to include a cover sheet?
- How much access do we have to this person?
- Are they so busy that we wouldn't be able to meet again for weeks?

These questions may have been answered by having worked with the CTO in the past, but we haven't. The last thing we want to do is spend time elaborating on this subject unnecessarily.

The diagram option

The most obvious and easiest solution to presenting our findings is a diagram, but as we've seen while establishing these links, a single diagram can become unwieldy. It might be visually pleasing if we create multiple diagrams, each showing an example of problem areas. In this exercise, in cases where more than one technology component is assigned to a technology service, we have a potential duplication of technologies. In those cases, we create a new diagram and drag the offending elements onto the diagram for illustration. The following figure shows an example of one potential problem area at *ABC Trading*:

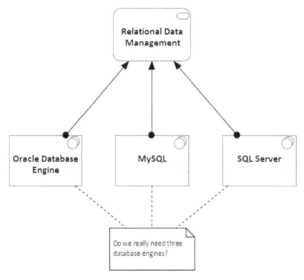

Figure 7.33 – Duplicate technologies

The report option

An alternative to a diagram is generating a report based on the contents of the repository. We can build a simple report template that produces a list in table form. The template is shown in the following figure:

```
package >
{Pkg.Name}
```

Service	Assigned Technology
element >	
{Element.Name}	connector > source > element > {Element.Name} < element < source < connector

```
< element
< package
```

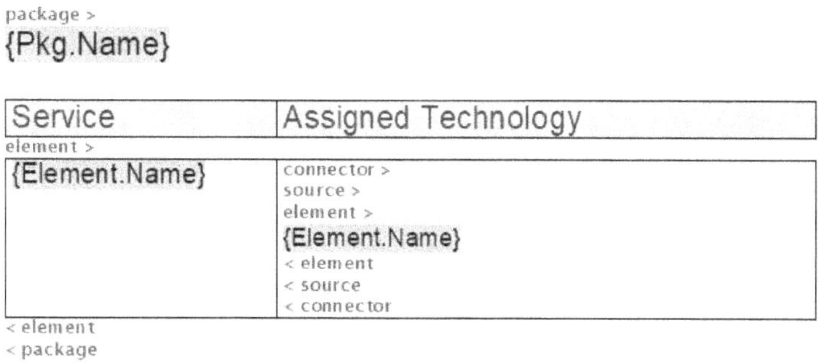

Figure 7.34 – A simple report template

We will learn more about building report templates in detail in *Chapter 11*, *Publishing Model Content*. That chapter will also show you how to include diagrams in reports and publish them to your audience. An example of report output is shown in the following figure:

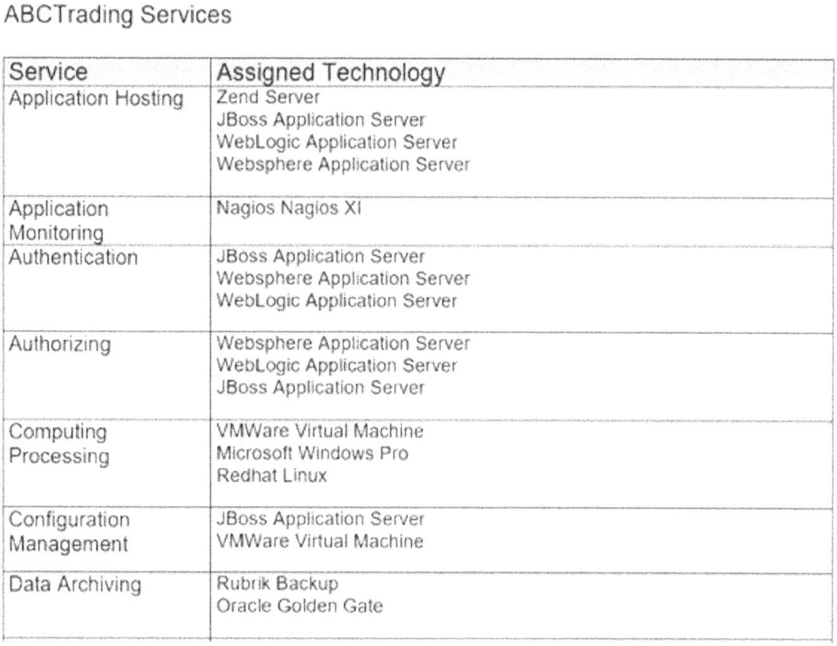

ABCTrading Services

Service	Assigned Technology
Application Hosting	Zend Server JBoss Application Server WebLogic Application Server Websphere Application Server
Application Monitoring	Nagios Nagios XI
Authentication	JBoss Application Server Websphere Application Server WebLogic Application Server
Authorizing	Websphere Application Server WebLogic Application Server JBoss Application Server
Computing Processing	VMWare Virtual Machine Microsoft Windows Pro Redhat Linux
Configuration Management	JBoss Application Server VMWare Virtual Machine
Data Archiving	Rubrik Backup Oracle Golden Gate

Figure 7.35 – A partial duplicate technologies report

As you can see, this report accomplishes the goal of identifying *potentially* duplicate technologies. Each service in the first column that contains more than one technology node in the second column can be a duplicate. For our scenario, we have decided to take an iterative approach. We will produce this simple report, then meet with the CTO, review the information, and determine at that time whether they need more information. The fact is that this information will need further scrutiny. In our scenario, however, this is sufficient to present to the new CTO. This may be all they need. More likely, they will solicit your help in narrowing this information down. That's fine, as all of this information is in reusable form; it's all in our repository. We can add, augment, or refine this information further as we see fit.

Summary

This was our first plunge into the technology behavioral elements of ArchiMate® 3.1. This was also our first time looking at an enterprise-wide set of information. We've covered a lot in this chapter, but we hope the principles didn't get lost in the detail. Our attempt is to be as practical as possible by reusing information that already exists.

We imported a large amount of information that was contained in spreadsheets, but that's not where the bulk of our effort was spent. It was actually much more difficult to identify the services provided by the technical nodes. However, once the services were mapped to nodes, the redundant technologies were easy to identify. While we are closer to reducing technology costs, by no means are we done. Much more work is needed to understand why the redundancies exist and how they can be reduced.

If you were an architect at *ABC Trading* for any length of time, you would probably already know where the redundancies exist and why. You may ask yourself why we can't just list those redundancies and be done with it. The answer is that this would be a much less effective method. Why? Because the information is not trusted.

Like it or not, information that comes from one person based on their personal knowledge is not trusted. We are human. We tend to include our own biases and misunderstandings, whether we intend to or not. If we are to move away from the all-so-common perception that architecture, especially enterprise architecture, lives in its own ivory tower, we need to present information in a way that lets others in the enterprise come to obvious conclusions on their own. We need to let our audience see what we already know. If we've done our job and presented the information correctly, they will come to the correct conclusions. They will also have gained trust in the enterprise architecture practice.

In the next chapter, we will introduce you to the business layer of the ArchiMate® language.

8
Business Architecture Models

The **business architecture** is the highest of the three core architecture layers in an enterprise. Many architects start their **Enterprise Architecture** (**EA**) practice at this layer, followed by the **application layer**, and then the **technology layer**.

The business architecture is often approached first because it traditionally occurs when the EA practice is defined and documented, with the intent to build the other layers around the business, which is a correct approach, and it keeps the business as the driver for changes. Additionally, TOGAF® has the business architecture as phase B of the **Architecture Development Method** (**ADM**), which guides its practitioners to have it in the early stages of the EA practice. However, due to the reasons that have been described in detail in *Chapter 1*, *Enterprise Architecture and Its Practicality*, one of the practical approaches that this book encourages is to start from any layer in the enterprise where there is a demand for EA input, and not to follow a waterfall sequence.

The business architecture layer is concerned with the business aspect of an enterprise, such as what the business provides to the world, how it achieves that, how the organization is structured, which business units are assigned to which functions, services, or processes, and how these business architecture elements are automated and realized by elements from the application and the technology layers. This is what we are going to cover in this chapter, and it can be summarized in two sections:

- Modeling the business structure
- Modeling the business behavior

If you are reading this book from the beginning, this chapter should be very easy and straightforward because you will have seen that most elements have been introduced in the application and the technology layers, as covered in the previous chapters. The only difference is that they cover a different aspect of the enterprise in this layer. If you are reading non-sequentially, then we recommend that you check out the following *Technical requirements* section and make sure that you have what you need to start.

Technical requirements

If you want to practice modeling while reading to achieve the best results, you need **Sparx Systems' Enterprise Architect**. If you do not have a licensed copy, you can download a fully functional 30-day trial version from the Sparx Systems website (`https://sparxsystems.com/products/ea/trial/request.html`).

We will continue adding the content of this chapter to the EA repository that we built in the previous chapters. If you want, you can download the repository file of this chapter from GitHub at `https://github.com/PacktPublishing/Practical-Model-Driven-Enterprise-Architecture/blob/main/Chapter08/EA%20Repository.eapx` instead of starting from scratch. Some of the steps in this chapter depend on elements that have been already created in the repository, so it is best not to start this chapter with an empty repository.

We will use the following ArchiMate® 3.1 specification chapters to guide our development:

- *Chapter 5, Relationships* (`https://pubs.opengroup.org/architecture/archimate3-doc/chap05.html#_Toc10045310`)
- *Chapter 8, Business Layer* (`https://pubs.opengroup.org/architecture/archimate3-doc/chap08.html#_Toc10045365`)
- *Chapter 12, Relationships Between Core Layers* (`https://pubs.opengroup.org/architecture/archimate3-doc/chap12.html#_Toc10045440`)

The metamodel diagrams in the aforementioned references will be used throughout this chapter, so we highly recommend that you print them and keep them within reach. When you are ready, we can start learning how to model the business structure.

Modeling the business structure

The business structure describes *what* the tangible elements are that compose the enterprise, and *who* is assigned to what. Modeling the business structure will answer questions such as the following:

- What are the business units that comprise this enterprise?
- What are the business roles that operate it, and who are the people assigned to these roles?
- Who are our customers and partners?
- What channels do these customers and partners use to access our services?

Now that we know the type of question we will answer in business structure models, let's start by defining the first business active structural element, which is the business actor.

Defining business actors

"A business actor represents a business entity that is capable of performing behavior." (https://pubs.opengroup.org/architecture/archimate3-doc/chap08. html#_Toc10045368)

Business actors can be internal or external to an enterprise. Customers and partners are considered external business actors, while business units and divisions within the enterprise are considered internal. **Business actors** within ArchiMate®'s context must not be confused with **actors** in UML. They both use the same *stickman* notation, and they are both called actors, which causes confusion. Actors in UML usually indicate users of the system that we are modeling, and these users can either be people or other systems. However, business actors in ArchiMate® represent business entities such as an entire organization, a division within it, or a person.

The more relevant elements to UML actors in ArchiMate® are business roles, which we will explain in more detail in the next subsection. To easily differentiate between business actors and business roles, always ask yourself, *is it a position?* If so, then it is a role. However, if it is a business unit or a specific person, then it is an actor. For example, *accounting* is a business unit, so it is a business actor. *Accountant* is a position, so it is a business role. *John Smith*, the accountant, is a business actor because individuals are business actors as well.

The last thing to keep in mind is that business actors can be assigned to multiple business roles, which makes sense as you often find some people playing multiple roles and occupying multiple positions in an organization.

ArchiMate® 3.1 provides two types of notation for modeling business actors, **Rectangular Notation** and **Borderless Notation**, as you can see in the following figure:

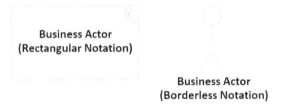

Figure 8.1 – Business actor notation

Business actors, in addition to other internal active structure elements, are responsible for providing the behavior of the business, such as business processes, business functions, and business interactions. We will talk in more detail about the business internal behavior elements in the *Modeling business behavior* section of this chapter. The following diagram shows the business actor-focused metamodel:

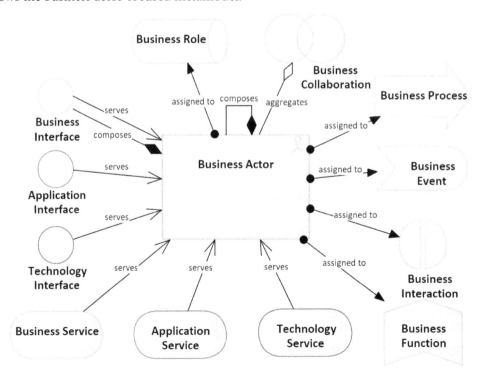

Figure 8.2 – The business actor focused metamodel

The simplest diagram that you can use business actors in is the organization chart diagram. You can show how the organization can be decomposed into business units that can themselves be decomposed into smaller business units, and so on. The following diagram shows an example of the **ABC Trading** organization chart:

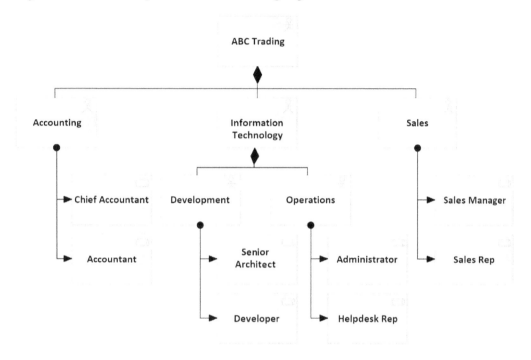

Figure 8.3 – The ABC Trading organization chart

We are sure that your actual organization chart will be larger than this example, so you may consider keeping the first level of the business actors in one diagram and having multiple child diagrams, each showing the details of a single business actor at a time. We will see more examples of modeling business actors as we explore the other business architecture elements, so let's read more about business roles first.

Defining business roles

"A business role represents the responsibility for performing specific behavior, to which an actor can be assigned, or the part an actor plays in a particular action or event." (https://pubs.opengroup.org/architecture/archimate3-doc/chap08. html#_Toc10045369)

Since business actors represent business units or business entities in general, **business roles** represent the actual responsibilities within the business units – in other words, the job positions that are supposed to perform the behavior. The accounting business unit requires specific roles, such as the chief accountant, controller, treasurer, and tax accountant. All these roles can be assigned to actual people who are business actors. Remember that business actors can represent specific people or business units, so a person can be assigned one or more roles within a business unit.

ArchiMate® 3.1 provides two types of notation for modeling business roles, **Rectangular Notation** and **Borderless Notation**, as you can see in the following diagram:

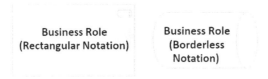

Figure 8.4 – Business role notations

Business roles are similar to UML actors. The only difference is that in UML, actors represent the users that perform actions (or behavior) on a specific system, while ArchiMate®'s business roles exist within the entire enterprise, not on a specific system. The following diagram shows the focused metamodel for the business role:

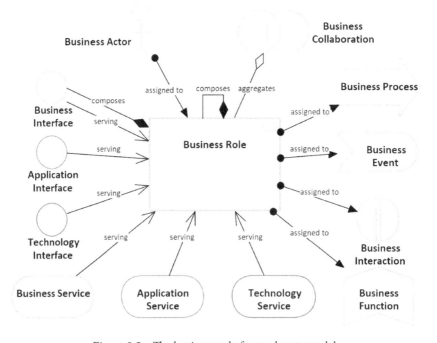

Figure 8.5 – The business role focused metamodel

As you can see, the business role focused metamodel is identical to the business actor-focused metamodel, and that's because they both specialize from the generic type of business internal active structure element and represent entities that are capable of performing behavior. Business behaviors such as business functions or business processes can either be performed by an entire business unit or by a specific role within that unit, and therefore, you can see that both business actors and business roles can be assigned to all business internal behavior elements.

A business actor can be assigned to one or more business roles, so in small enterprises, you can still use an organization chart from a large enterprise, but in this case, you will require your business actors to play multiple roles, which is a very common practice. The opposite is also true – you can have multiple actors assigned to the same role. It all depends on the size of your enterprise and the skill sets that your business actors possess. The following diagram shows an example of how business actors can be assigned to business roles:

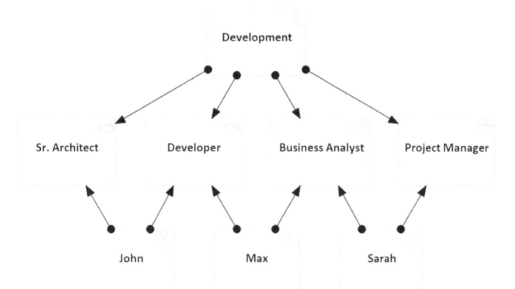

Figure 8.6 – Business actors assigned to business roles

One of the very powerful features of Sparx is that it can tell you how many items are linked or connected to a specific item. If you close the preceding diagram or even delete it, you will still have the items and the defined relationships in the repository. Right-clicking on the **Business Analyst** role in the project browser and choosing **Properties** > **Properties** > **Links** from the context menu will show you all the items linked to it and the types of the relationships. This way, you can find out that *Max* and *Sarah* are both assigned to this role even if you do not see the diagram.

Each business role is expected to perform a specific set of behaviors, such as business functions, business processes, and business interactions. The processes that a project manager is expected to perform can be described in a diagram such as the following:

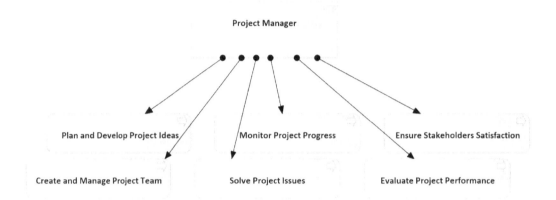

Figure 8.7 – A business role assigned to business functions

One of the most beautiful things about ArchiMate® is that it allows you to describe almost everything in an enterprise in a model. For example, the preceding diagram can be used as a job description for the **Project Manager** role. You can develop a diagram like this for each business role that you have within your organization. The human resources department can publish this information on the company's website or intranet as part of its documentation process.

We will learn more about publishing content from Sparx in *Chapter 11, Publishing Model Content*. For now, let's look at another business structural element, which is business collaboration.

Defining business collaboration

"A business collaboration represents an aggregate of two or more business internal active structure elements that work together to perform collective behavior." (https://pubs.opengroup.org/architecture/archimate3-doc/chap08.html#_Toc10045370)

A **business collaboration** element can be an aggregate of multiple business actors or business roles to provide a collaborative behavior. A joint venture between a shipping firm and an online shopping business is an example of business collaboration. It is a strong form of partnership but weaker than merger and acquisition. Both businesses can still act individually and independently from each other, but together, they can behave differently.

ArchiMate® 3.1 provides two types of notation to model business collaboration elements, as you can see in the following diagram:

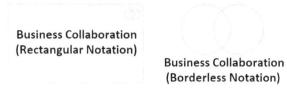

Figure 8.8 – Business collaboration notation

The business collaboration-focused metamodel is a replica of the business actor- and business role-focused metamodels, as you can see in the following diagram:

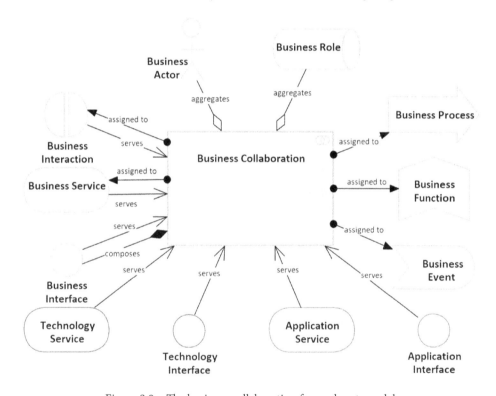

Figure 8.9 – The business collaboration focused metamodel

If *ABC Trading* and one of its partners decided to increase the level of partnership to form a union, that is a business collaboration. It is a form of aggregation that is stronger than a partnership but weaker than a merger. Through this collaboration, both partners will be able to provide services and perform processes in ways that would have been impossible individually.

The next structural element that we will look at is the business interface.

Defining business interfaces

"A business interface represents a point of access where a business service is made available to the environment." (https://pubs.opengroup.org/architecture/archimate3-doc/chap08.html#_Toc10045371)

Businesses must provide services in order to generate revenue and remain operational. Business services are provided through interfaces, and these interfaces can take the form of a web portal, a mobile app, a kiosk, a customer service number, or a reception office inside your physical location. Therefore, business interfaces are not necessarily computerized or automated. Digital business services need to be provided through digital business interfaces for sure, but if your business fixes cars, then it must be provided in a workshop. The reception office in the workshop in this case is the business interface. Think of business interfaces as the *channels* that make services available to their consumers.

ArchiMate® 3.1 uses two types of notation for modeling business interfaces, as you can see in the following diagram:

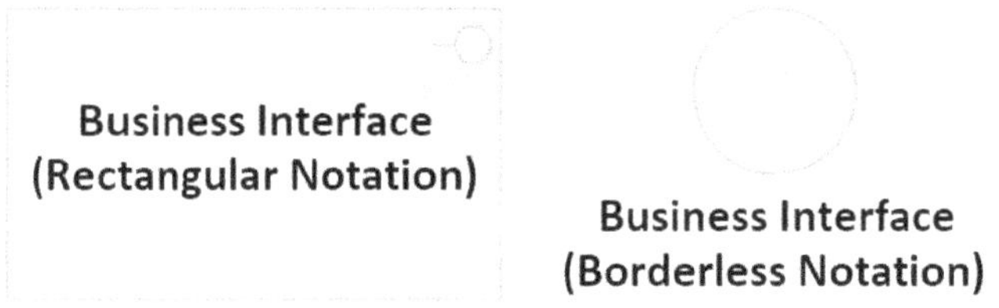

Figure 8.10 – Business interface notation

Business interfaces provide services to structural components in the business, application, and technology layers. The following diagram shows the focused metamodel for the **Business Interface** element:

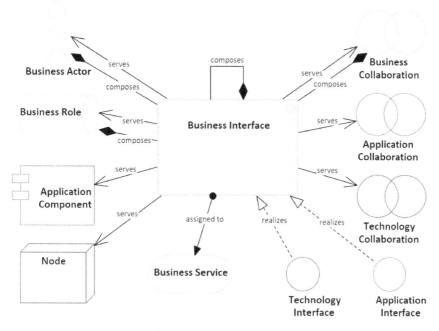

Figure 8.11 – The business interface focused metamodel

An automated (or digitized) business interface is one that has been realized by one or more application or technology interfaces. If it is not realized by any, it means that it is a manual or *in-person* business interface. Some types of services must be provided in person, no matter how advanced your business is. The following diagram shows an example of how multiple business interfaces are assigned to one business service:

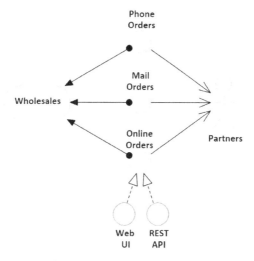

Figure 8.12 – Multiple business interfaces assigned to a service

The preceding diagram tells us that there are three business interfaces assigned to the **Wholesales** business service. One of them is digitized and is realized by two application interfaces. It is also possible to assign the same business interface to multiple business services, which means that this interface is used to provide more than one service to the consumers.

The next and last structural element that we will explore is the business object.

Defining business objects

"A business object represents a concept used within a particular business domain."
(`https://pubs.opengroup.org/architecture/archimate3-doc/chap08.html#_Toc10045381`)

Business objects represent information that means something within a specific business domain. Business objects such as customer, invoice, package, and product all have meaning within the *ABC Trading* company, while claim, deductible, encounter, and policy are information that has meaning within another business that deals with insurance. Regardless of whether these business objects are digitized or not, and regardless of the digital format they exist in, they are considered business objects within the business context. Digitized business objects are the ones that are realized by data objects and technology objects.

Let's talk a little more about this to clarify it. The product, for example, is the physical item that you sell to your customers, and the information that you know about it is the product's business object. A customer is a business actor that consumes services from your business. The information that is related to the customer actor, such as name, phone number, and address, is the customer business object, so we're talking about two architectural elements here – the customer actor and the customer business object.

Whether the information about the customer business object is stored in paper files, in the heads of the salespeople, in databases in binary, or even in quantum formats, they are all different forms of realizing the same business object by different technologies. Even paper files and cabinets, while outdated, are considered technologies. However, the *concept* of the customer within the context of the business is always the same, and that is what business objects represent.

Unlike most other elements, ArchiMate® 3.1 provides only one type of notation for modeling business objects. The following diagram shows the business object focused metamodel with the **Business Object** notation in the middle of it:

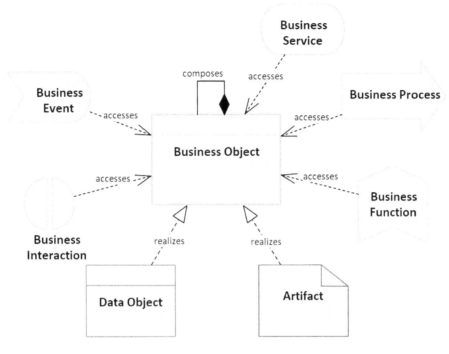

Figure 8.13 – The business object focused metamodel

Because business objects are passive elements, the only permitted relationship between them and the other business architecture elements is the *access* relationship. The following diagram shows an example of how business process elements access business objects:

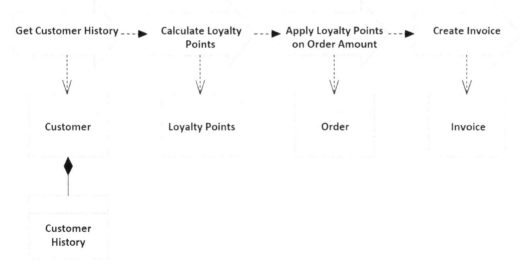

Figure 8.14 – Business processes accessing business objects

In real-world examples, your objects will be architected differently, such as having the customer history as an independent element or having loyalty points as part of the customer profile. It depends on how your business recognizes the objects; the size does not matter. If loyalty points mean something to the business, then you may model them in their own business object, as we did. If they are meaningless without being part of a bigger business object, then you may go with a different design. Either way, the concept of what can be considered a business object is clear, we hope.

This concludes our section about modeling business structure, and since only describing the structure of the business is not enough, we need to describe the behavior as well, and that is what we will be covering in the next section.

Modeling business behavior

Business behavior models show how your business performs things such as the following:

- How services are provided
- How requests are handled internally
- What are the steps to be followed, and are they automated or not?
- What resources are required to perform which behavior?

All these questions and more can be answered in architecture models that describe business behavior. This is what we will learn in this section, and we will cover these topics:

- Defining business services
- Defining business processes
- Defining business functions
- Defining business interactions
- Defining business events

With that said, let's start with the first business behavior element – the business service.

Defining business services

"A business service represents explicitly defined behavior that a business role, business actor, or business collaboration exposes to its environment." (`https://pubs.opengroup.org/architecture/archimate3-doc/chap08.html#_Toc10045378`)

Business services in general are the visible parts of the business that are exposed to the external environment. A customer does not see what business units your organization has, nor which actor is assigned to which role. They also do not see how you work internally to deliver the services. All that they see and know about your business are the exposed parts, which are the business services.

However, the concept of business services does not only apply to the entire enterprise and its external service consumers. Each internal business unit can be perceived as a provider of services to other business units within the same enterprise. Additionally, each business role can be assigned to provide a specific set of services to other business roles and business actors. Therefore, the business services are contextually defined by the business structural element that is assigned to them.

ArchiMate® provides two types of notation for modeling business services, **Rectangular Notation** and **Borderless Notation**, as you can see in the following diagram:

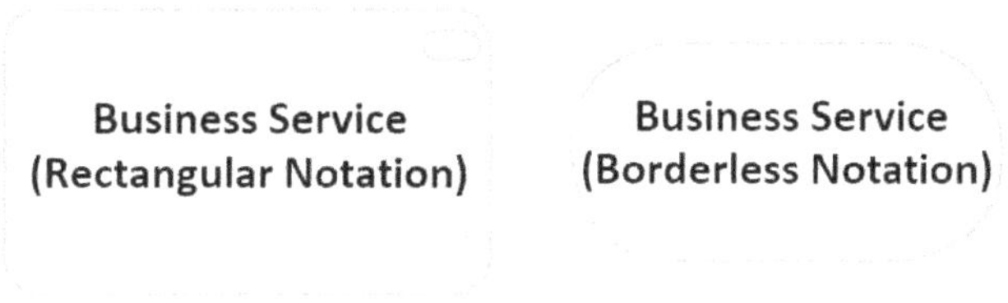

Business Service
(Rectangular Notation)

Business Service
(Borderless Notation)

Figure 8.15 – Business service notation

Because the business services are meant to be exposed to serve other elements, this metamodel is busier than other focused metamodels, as you can see in the following diagram:

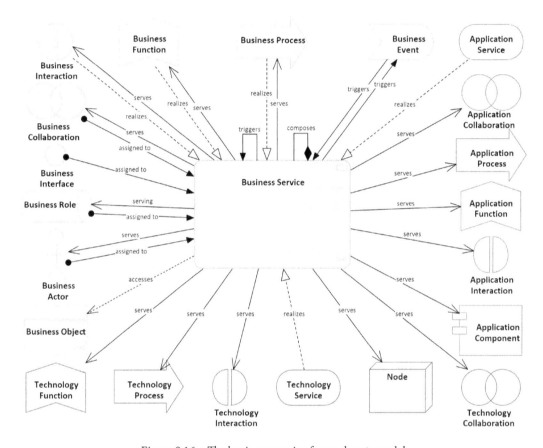

Figure 8.16 – The business service focused metamodel

Let's take examples to clarify some ideas on business services and the preceding metamodel. As mentioned earlier in this subsection, business services that are provided by an entire enterprise can vary from business services that are provided by a specific business unit inside that enterprise. Partners of *ABC Trading* can only see what it offers outside its environment, as depicted in the following diagram:

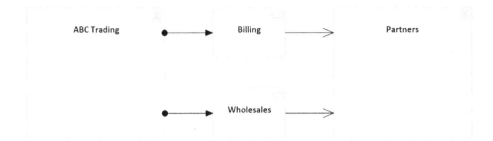

Figure 8.17 – ABC Trading exposed services

However, with an inside look at *ABC Trading*, we can see different sets of services that are serving different sets of consumers, as indicated in the following diagram. The **Human Resources** internal business actor (or business unit) is assigned to the **Recruiting** service, which is an internal service that only other internal business actors can see and use. **Accounting**, on the other hand, is assigned to two business services – **Billing** and **Salary Deductions**. The **Billing** service serves external partners, while the **Salary Deductions** service serves the **Human Resources** internal business actor, as shown here:

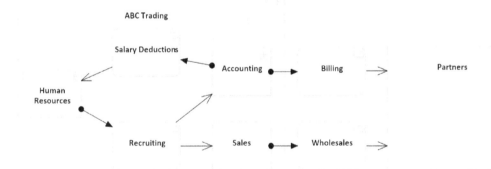

Figure 8.18 – Different services provided within a different scope

External business actors such as partners do not see the internal business services. Moreover, they are also not concerned about which business unit provides the services. They see that the *ABC Trading* business actor is assigned to the billing and wholesales services.

A business services catalog is an architectural artifact that contains a list of all the business services that are provided by a specific business actor (or business role or business collaboration) to its external environment. So, whenever you develop a business services catalog, you need to ask who the business actor is that you are modeling, what is considered internal to it, and what is considered external.

Then, only consider the services that target its external environment. Only consider what the external environment sees from this business actor. Recruiting, for example, in *Figure 8.18* cannot be considered part of the *ABC Trading* business services catalog because it is an internal service. However, if you are modeling the human resources business actor and want to define its service catalog, then recruiting will be a part of it for sure.

The next behavior element that we need to look at closely is the business process.

Defining business processes

"A business process represents a sequence of business behaviors that achieves a specific result such as a defined set of products or business services." (https://pubs.opengroup. org/architecture/archimate3-doc/chap08.html#_Toc10045374)

For a business service to be realized and provided to customers, the business needs to perform some activities in a specific sequence. This sequence is what is referred to as **business processes**. The maturity of any business can be measured by how well documented its business processes are. Business processes need business structural elements to be assigned to them because they need something or someone to perform them. They also need to have access to a certain type of information in order to perform their activities.

Business processes exist in every business all the time. Building the pyramids must have followed a specific process or else they wouldn't exist. Building a colony on Mars would also follow a specific process. Automated or not, that's a different story, but business processes have no relation to technology. Technology makes them faster and more efficient, but the automated processes are defined using the application process and the technology process elements. They realize the business processes or parts of them. A fully automated business process is one that has all its activities realized by an application process or a technology process.

ArchiMate® provides two types of notation for modeling business processes, as you can see in the following diagram:

Figure 8.19 – Business process notation

Business processes can aggregate business functions, and business functions can aggregate business processes. In most cases, business functions are at a higher abstraction level than processes, so they aggregate them. In some cases, you may find that high-level business processes aggregate mid-level business functions, which aggregate lower-level business processes. There is no limit to the deepness of this nesting loop, but too much nesting can become confusing, so keep that in mind.

A quick and easy way to differentiate the two is that business functions are abstracted and do not have a sequence, while business services have more details and contain a sequence. We will talk in more detail about business functions in the next subsection, *Defining business functions*, with more examples of how to differentiate the two. The following diagram shows the **Business Process**-focused metamodel:

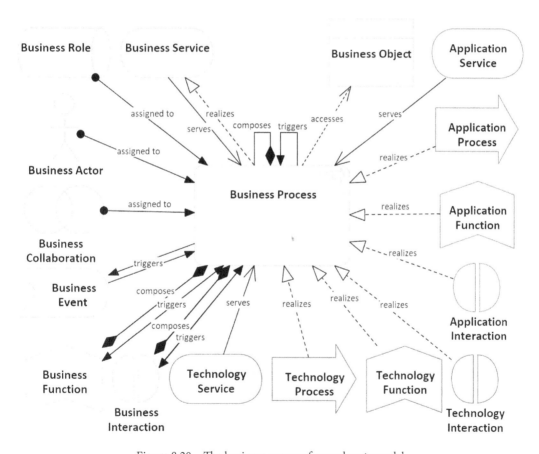

Figure 8.20 – The business process focused metamodel

Business processes can be triggered by any other behavioral element, such as business events, business functions, and business interactions. They can also be triggered by other business processes. When one process finishes, it triggers the next process, and the control flows from one process to another until the entire process is complete, as shown in the following diagram:

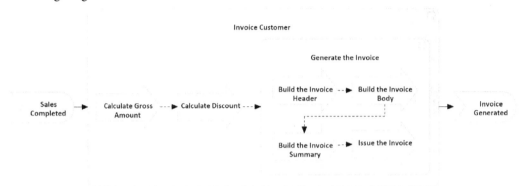

Figure 8.21 – A multi-level business process

When a business process is automated, it means that it has been realized by one or more application and technology processes. A single business process can have different application processes realizing it in the same application or different applications, each realizing a part of it. If you are transitioning from a legacy application to a modern one, there is a high probability that you will find that the same business process is realized more than once in multiple applications. The following diagram shows an example of a business process that is realized by multiple application and technology processes:

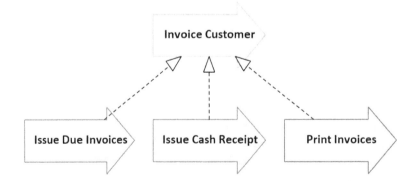

Figure 8.22 – A business process realized by application and technology processes

Remember not to confuse business processes with application processes and technology processes. They are all types of processes, so they all have a sequence and aim to achieve specific results. They are just describing other layers of the enterprise. Please refer to the *Describing application behavior* section of *Chapter 5, Advanced Application Architecture Modeling*, to learn more about application processes. Also refer to the *Using technology behavioral elements* section of *Chapter 7, Enterprise-Level Technology Architecture Models*, to learn more about technology processes.

The next behavioral element that we will explore is the business function, which is tightly related to the business process. Let's see how.

Defining business functions

"A business function represents a collection of business behavior based on a chosen set of criteria (typically required business resources and/or competencies), closely aligned to an organization, but not necessarily explicitly governed by the organization." (https://pubs.opengroup.org/architecture/archimate3-doc/chap08.html#_Toc10045375)

Business functions are internal behavioral elements, so they are only visible internally, inside the enterprise. They can compose or be composed of other behavioral elements, such as business processes, business interactions, and other business functions. So, a business function can be composed of business processes and business functions. Business processes can be composed of other business processes and business functions.

Business functions are abstracted and answer the question of what the business performs internally. Business functions do not stipulate how to do things, so they do not involve a sequence of actions. Business functions can be grouped based on criteria that the business decides. It could be geographical, financial, managerial, or any other grouping that makes sense to the business. You can have eastern region sales and western region sales, for example, as two business functions under a bigger business function called sales. The two regional functions can share some resources, but they also have their own resources assigned to them. They may also perform things differently, which means that they comprise different sets of business processes and/or lower business functions.

Business functions need resources such as business actors and business roles to be assigned to them because behavior cannot be performed without a structure. This is why business units (actors) and business functions can carry the same names. The human resources business unit (actor) is assigned to the human resources business function. They are two elements despite having the same or similar names. The actor can either be internal or external to the enterprise, so some business functions can be performed by third parties. The customer service function is a very common example of an outsourced business function.

ArchiMate® 3.1 provides two types of notation for modeling business functions, as you can see in the following diagram:

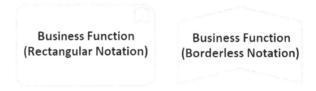

Figure 8.23 – Business function notation

In the following diagram, you can see the focused metamodel for the **Business Function** element. Because business functions and business processes are both internal behavioral elements, their focused metamodels are almost identical:

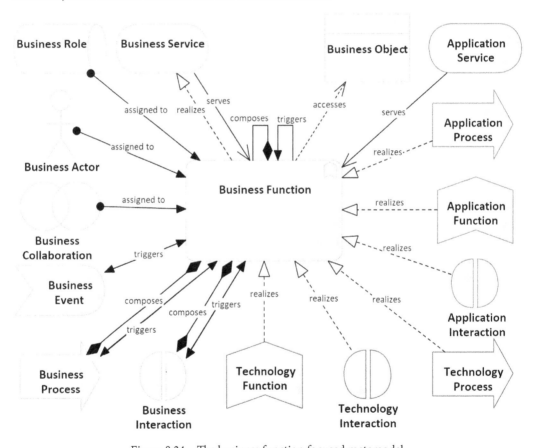

Figure 8.24 – The business function focused metamodel

Business functions are realized by application functions and technology functions. The following diagram shows you an example:

Figure 8.25 – A business function realization

As we can see, the **Human Resources** business actor is assigned to the **Human Resources Management** business function. In this company, they use a business application for managing enterprise resources, and one of its application functions is **Human Resources Management**. Even with the business function and the application function are having the same name, they are still two architectural elements in two different layers.

The following diagram shows how a business function can be composed of smaller business functions, which themselves can be composed of smaller business functions or business processes:

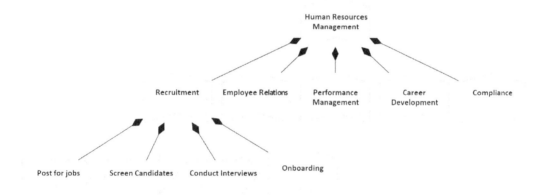

Figure 8.26 – An example of the human resources business function

Using the preceding diagram as a reference, you can create your enterprise business functions catalog. Do not be surprised if you end up with a diagram that is very close to your organization chart because, in many organizations, business units are assigned to business functions in a one-to-one relationship.

The next behavioral element on our list is the business interaction element.

Defining business interactions

"A business interaction represents a unit of collective business behavior performed by (a collaboration of) two or more business actors, business roles, or business collaborations." (`https://pubs.opengroup.org/architecture/archimate3-doc/chap08.html#_Toc10045376`)

Business interactions are just like other behavior elements (business services, business processes, and business functions) – they describe a behavior that is performed by a business collaboration, not a single business actor or role. Therefore, business interactions exist within your enterprise only when there are business collaborations.

Business interactions can be composed of multiple business functions and processes, which means that the business interactions that are performed by a business collaboration element are actually composed of business functions and processes that are performed by the structural elements (actors and roles) composing the business collaboration.

ArchiMate® 3.1 provides two types of notation for modeling business interactions, as you can see in the following diagram:

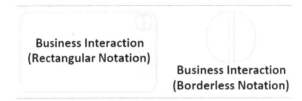

Figure 8.27 – Business interaction notation

Due to the limited usage of business interactions, we will leave it to you to develop their focused metamodel, which will be identical to the metamodels for **Business Process** in *Figure 8.20* and **Business Function** in *Figure 8.24*.

The last behavioral element that we will look at is the business event element.

Defining business events

"A business event represents an organizational state change." (https://pubs.
opengroup.org/architecture/archimate3-doc/chap08.html#_
Toc10045377)

Business events occur either internally within an enterprise or externally to it. When they occur, they trigger other behavioral elements, such as business processes, functions, interactions, and services. An external event can be like a customer submitting a request, a partner who is out of business, a change in the supply or demand of a specific product, a merger, or an acquisition of a new business. Examples of internal events include the establishment of a new business role, the closure of a business unit, the reception of a customer request, the fulfillment of a customer request, the completion of end-of-month payroll processing, and achieving the quarterly financial goals.

Just like many other elements, ArchiMate® 3.1 provides two types of notation for modeling business events, as you can see in the following diagram:

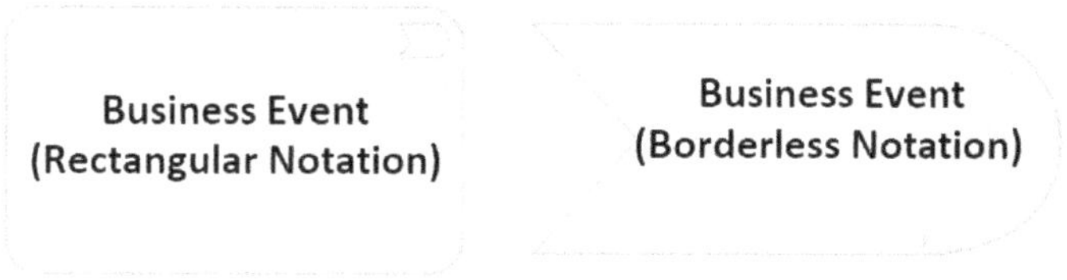

**Business Event
(Rectangular Notation)**

**Business Event
(Borderless Notation)**

Figure 8.28 – Business event notation

Business events trigger and get triggered by other business behavioral elements. The submission of a customer request will trigger a series of business processes to fulfill that request. These processes can themselves trigger other behavioral elements, such as other business processes, functions, interactions, and events. This series of actions will keep going until there is no more behavior to be triggered. *Figure 8.29* shows an example of a business event triggering a business process that contains smaller business processes, and finally, it triggers another business event upon completion.

Additionally, and just like all other behavioral elements, business events need a business internal active structure element to be assigned to them, such as a business actor, a role, or a collaboration. The following diagram shows the focused metamodel of business events:

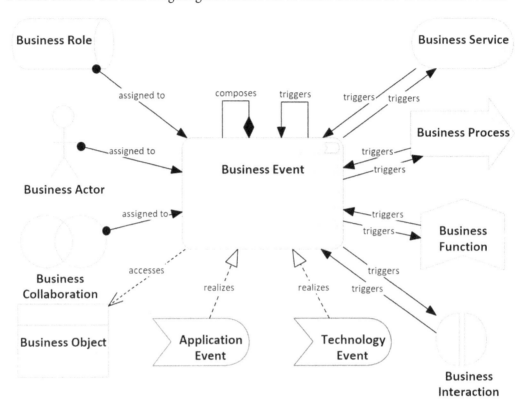

Figure 8.29 – A business event focused metamodel

Automated business events are realized by application and technology events, but not all business events are necessarily automated. Customers' requests can still be received over the phone or by mail, for example. If the organization has an application that allows customers to submit requests online, then an event inside that application will be fired and trigger a series of automated application processes to automatically fulfill the request. The following example depicts how automated business events and business processes can be realized:

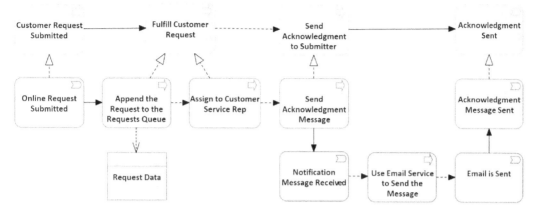

Figure 8.30 – An example of realized business events and processes

This diagram shows that to completely automate business events and processes, you may need to realize them through multiple elements in multiple architecture layers, and that realization is not necessarily one-to-one. Some business processes, as you can see, get realized by one or more application and technology processes.

We hope that this chapter has helped you understand the principles of modeling business behavior; let's summarize what we have learned.

Summary

As we have seen in this chapter, business architecture elements look very similar to their counterpart elements in the application and technology layers. However, even if they look alike, the business architecture elements describe the business regardless of the availability of applications and technology.

Automated business elements are realized by application and technology elements. Knowing this is critical for many businesses as they seek higher maturity levels. If your organization keeps everything in an EA repository, as it should, elements can be located in a few minutes. This is the power of EA, and this is the helpfulness of an EA tool such as Sparx.

In the next chapter, we will combine the knowledge and the experience that we have gained regarding building business, application, and technology models under the business capabilities umbrella.

9
Modeling Strategy and Implementation

The **strategy layer** is a layer on top of the **business**, **application**, and **technology** layers. Because everything in the enterprise must happen to realize the strategy, having strategy elements within the repository that are supported with well-designed models can tell management which enterprise elements are helping to realize the strategy.

It will help management to predict what would happen if changes were made to the state of an organization. Changes to the state of the organization can be any changes that require the business to respond, and they can be positive or negative changes. They can be financial, political, natural, or technological. Some changes bring opportunities and others bring threats, and your business must be ready to respond. Otherwise, it will be in danger of being left behind.

Business capabilities are at the heart of strategic planning and modeling. Modeling business capabilities is one of the most powerful tools that can help organizations to document how they are able to provide their services to customers, and how they can build on their abilities if they don't have them.

In this chapter, you will learn how to utilize the power of the strategy models and how to use them to identify the gaps between business today and business in the future. In architecture words, this chapter will cover the following two topics:

- Introducing strategy elements
- Introducing implementation elements

But before you move on to the first topic, make sure that you have what is listed in the *Technical requirements* section.

Technical requirements

It is always better to practice modeling while reading, and better results will be achieved. If you do not have a licensed copy of **Sparx Systems' Enterprise Architect**, you can download a fully functional 30-day trial version from the Sparx Systems website (`https://sparxsystems.com/products/ea/trial/request.html`).

We will continue adding the content of this chapter to the EA repository that we built in the previous chapters. If you want, you can download the repository file of this chapter from GitHub at `https://github.com/PacktPublishing/Practical-Model-Driven-Enterprise-Architecture/blob/main/Chapter09/EA%20Repository.eapx`, instead of starting from scratch. Some of the steps in this chapter depend on elements that have been already created in the repository, so it is better to not start this chapter with an empty repository.

We will use the following **ArchiMate® 3.1** specification chapters to guide our development:

- *Chapter 5, Relationships* (`https://pubs.opengroup.org/architecture/archimate3-doc/chap05.html#_Toc10045310`)

- *Chapter 7, Strategy Elements* (`https://pubs.opengroup.org/architecture/archimate3-doc/chap07.html#_Toc10045354`)

- *Chapter 13, Implementation and Migration Elements* (`https://pubs.opengroup.org/architecture/archimate3-doc/chap13.html#_Toc10045444`)

The metamodel diagrams in the aforementioned references will be used throughout this chapter, so we highly advise you to print them and keep them in reach for maximum benefit. When you are ready, we can start learning how to model the business strategy.

Introducing strategy elements

The strategy layer uses elements for modeling strategy artifacts, just like all the other layers. Strategy elements are also of two types, **structural** and **behavioral**, just like the elements in other layers. In this section, we will explore the four elements of the strategy layer, which are as follows:

- Capabilities
- Value streams
- Courses of action
- Resources

As we learn about these elements, we will elaborate on each one, with examples to help you understand how to use them to develop strategy models in your work environment. We will start with capabilities.

Defining capabilities

"A capability represents an ability that an active structure element, such as an organization, person, or system, possesses" (`https://pubs.opengroup.org/architecture/ archimate3-doc/chap07.html#_Toc10045359`).

The topic of **business capabilities** is one of the most controversial topics in **Enterprise Architecture** (**EA**). I always say that if you are sitting in a room full of enterprise architects and don't have anything to talk about, ask the question, *what are business capabilities?* Then, I guarantee you an endless debate that will take forever without an agreement. One of the reasons behind this huge disagreement among enterprise architects is a result of the vague definitions that are provided by the different standards. If the people who are writing the standards are unable to agree on a clear definition, then don't blame the architects who are following these standards. Try it yourself and search the internet for the meaning of business capabilities; you will see exactly what I mean.

In this section, we are introducing you to our understanding of business capabilities, which are based on ArchiMate®'s definition and our long practical experience in EA. Always keep in mind that what matters the most, as a practical enterprise architect, is to know how to model business capabilities and present them to management rather than keep arguing about them.

ArchiMate®'s definition of a capability is also very abstracted and does not answer any of the questions that enterprise architects will be looking for, so we will try to elaborate on it. To start with, a **capability** is a strategic behavior element, which means it is an intangible element. It is a behavior that an active structural element can have or own (*possess*). Owning this behavior enables the structural element to be *capable* of realizing a business outcome by providing value. This value can take the form of services that can be provided or products that can be sold.

> **Important Note**
>
> Sparx combined elements from the strategy and the motivation layer into one toolbox, which is called **motivation**. If you want to create a strategy model, you need to choose motivation as the type of the diagram.

ArchiMate® 3.1 provides two notations for modeling capabilities, as you can see in the following diagram:

Figure 9.1 – Capability modeling notations

The following focused metamodel shows how every core behavioral element in the enterprise realizes the capability:

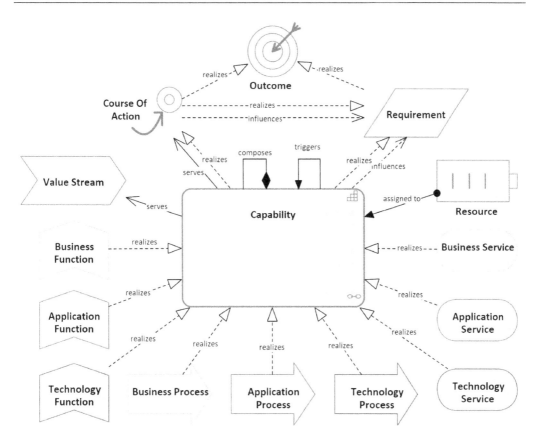

Figure 9.2 – The capability focused metamodel

As you can see, every function, service, and process, from the business, application, and technology layers, *realizes* the capability. In other words, every enterprise behavior element *realizes* the capability. We can confidently say that capabilities are the center of the enterprise, and this statement is made by us, not by any standard. We have already introduced the behavioral elements from the application layer in *Chapter 5, Advanced Application Architecture Modeling*, and we also introduced the behavioral elements from the technology layer in *Chapter 7, Enterprise-Level Technology Architecture Models*. We also introduced the behavioral elements from the business layer in *Chapter 8, Business Architecture Models*. Please refer to these chapters if you want to refresh your memory on these elements.

There are a few things to notice in the previously focused on metamodel in *Figure 9.2*. The first thing is that capabilities realize and influence requirements. If a car manufacturer, for example, decided to build motorcycles, there must be some strategic requirements behind this decision. This requires the business to develop the ability to make new products and provide new services.

When the required capability is developed, the business will be able to achieve (or realize) the goals. At the same time, requirements can be influenced by existing and targeted business capabilities. The decision for a car manufacturer to build motorcycles, not computers, for example, could have been influenced by the existing capability of manufacturing cars, which are relatively closer to motorcycles than to computers. In general, our requirements have always been influenced by our possessed capabilities, even at a personal level. We all want to have everything, but what limits us are our capabilities.

Another thing to notice is that the capability realizes and serves the course of action. A course of action is simply the plan that you need to follow to achieve the desired outcome based on the capabilities that you have. We will talk more about the course of action element in the *Courses of action* section, later in this section. But in a nutshell, if you have the capability but you do not have the proper plan to achieve your desired outcome, that capability will be wasted. Therefore, capabilities need to realize the courses of action needed to actualize (realize and influence) business outcomes.

Remember that capabilities are meant to deliver value, so let's understand what value and value streams are.

Value streams

"A value stream represents a sequence of activities that create an overall result for a customer, stakeholder, or end user" (`https://pubs.opengroup.org/architecture/ archimate3-doc/chap07.html#_Toc10045360`).

Value is anything that a stakeholder wants to have from an organization. Value from the perspective of a customer takes the form of the products or services that the organization offers, while value for a shareholder takes the form of profit, market share, reputation, and customer loyalty. The products that the organization produces may not be of interest to the owners as they look for different value.

Some values are dependent on other values, which in turn depend on other values. This is what is known as a **value stream**. It is a sequence of values, activities, and actions that together participate in realizing an outcome. Do not confuse value streams with processes, as they both involve a sequence of actions. Value streams represent a strategic high-level sequence of actions, while processes represent an operational-level sequence of actions. They are both behavioral elements, and processes need ultimately to realize the value streams, but each has a different scope of coverage and different level of detail.

ArchiMate® 3.1 provides two notations for modeling value stream elements, as you can see in the following diagram:

Figure 9.3 – Value Stream modeling notations

Value streams run if the organization can do so or, in other words, is capable of doing so. To deliver value, you must have the right capability. In architectural terms, we can say that capabilities serve value streams.

Since value streams are strategic behavioral elements, they are realized by other behavioral elements, such as processes, functions, and services from the business, application, and technology layers. The following diagram shows the focused metamodel for value streams:

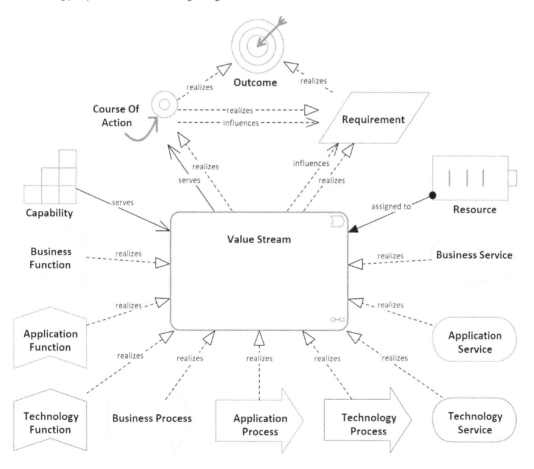

Figure 9.4 – The Value Stream-focused metamodel

As you can see, it is identical to the capability-focused metamodel because both value streams and capabilities are strategic behavior elements. You can also see that value streams need resources assigned to them, just like capabilities. Without resources, such as actors, application components, technology components, data, information, technology components, and other structural elements, no behavior will be performed. Talking about resources, let's learn more about them.

Defining resources

"A resource represents an asset owned or controlled by an individual or organization" (https://pubs.opengroup.org/architecture/archimate3-doc/chap07. html#_Toc10045357).

As you can see again, the definition is very vague and does not answer many of the questions that you may have about capabilities. As mentioned in the previous *Defining capabilities* subsection, a capability is a behavior that needs people, information, and technologies. All of these are forms of resources, so we must assign a specific number of resources to be able to deliver a specific capability.

ArchiMate® 3.1 provides two notations to model resource elements, as you can see in the following diagram:

Figure 9.5 – Resource modeling notations

The following focused metamodel shows how resources are assigned to capabilities and value streams, and they all realize and influence requirements:

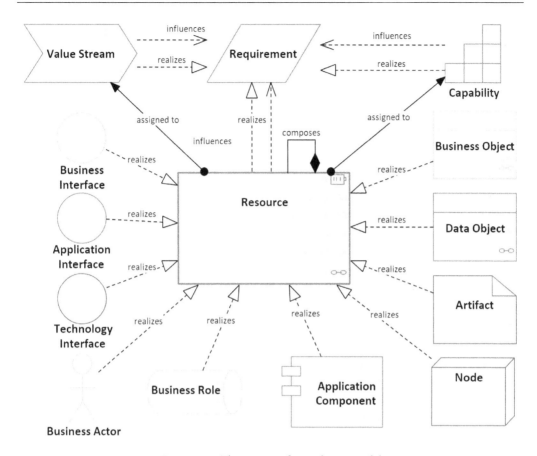

Figure 9.6 – The resource focused metamodel

Resources are structural elements, and they are realized by the active structural elements in the business, application, and technology layers. We have already introduced the structural elements from the application layer in *Chapter 5, Advanced Application Architecture Modeling*, introduced the structural elements from the technology layer in *Chapter 6, Modeling in the Technology Layer*, and we also introduced the behavioral elements from the business layer in *Chapter 8, Business Architecture Models*.

Resources in the project management world include time and money in addition to human resources and physical assets. Time and money are not part of the ArchiMate® specification at any layer, so you cannot model them. However, you can still model what they can buy (or hire) for you. So, you can model the human resources that you can hire, the application and technology components that you can buy, and the information that you can gain with time and money.

The last strategy element that we will explore before giving you some practical examples to model strategy elements is the course of action element.

Courses of action

"A course of action represents an approach or plan for configuring some capabilities and resources of the enterprise, undertaken to achieve a goal" (`https://pubs.opengroup. org/architecture/archimate3-doc/chap07.html#_Toc10045361`).

Organizations, in general, have multiple capabilities. These capabilities can be grouped and organized in different ways to achieve different outcomes. The courses of action is where you plan and configure this grouping of capabilities and value streams based on the desired outcome. You can think of courses of action as long-term strategic initiatives rather than thinking of them as short-term detailed plans. They just outline what needs to be done without detailing how.

Take, for example, the goal of sending people to Mars. To achieve this goal, one of your courses of action would be to build a spaceship that is capable of landing on Mars and returning. The details on how to develop the needed business capabilities to build that spaceship are at a more tactical short-term level.

Keep in mind that courses of action can compose larger courses of action and can be composed of smaller courses of action. So, being at a high level of abstraction does not mean that they cannot be part of some other higher-level or lower-level strategic initiatives.

ArchiMate® 3.1 has two notations to model course of action elements, as you can see in the following diagram:

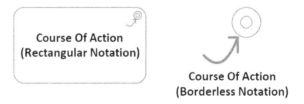

Figure 9.7 – Course of action modeling notations

courses of action are strategy behavioral elements, so they are realized by capabilities and value streams, as you can see in the following focused metamodel:

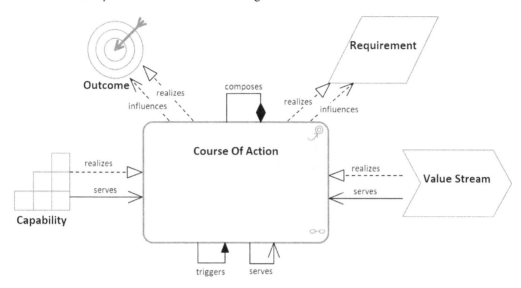

Figure 9.8 – The Course of Action focused metamodel

Courses of action can be composed of smaller courses of action, and they can serve, be served, trigger, or be triggered by other courses of action.

Let's look at some examples to illustrate the relationships between the strategic structural elements, the strategic behavioral elements, and the structural and behavioral elements from other enterprise layers.

Modeling the ABC Trading strategy

ABC Trading has decided to establish online retail sales to sell goods directly to customers and reduce its dependency on retail partners. According to feasibility studies that *ABC Trading* has conducted, the CEO has decided to proceed with the initiative and asked you to develop a high-level business architecture document. You need to help the CEO to model their thoughts, describe the targeted online service, and cooperatively identify what *ABC Trading* needs to do to provide the service. In this subsection, we will be doing the following:

- Modeling the value stream
- Realizing capabilities by behavioral elements
- Differentiating between the **as-is** and the **to-be** elements
- Identifying the needed resources

We will start by modeling the value stream and the capabilities that serve it.

Modeling the value stream

As mentioned previously, first things first. Before we start modeling capabilities in detail, it is best to start by answering the question of *why* we need them. The best person who can give these answers is the CEO or manager who assigns the building and development of those capabilities. It is very important to understand that as an enterprise architect, you are not the one who is finding these answers. Your role is to capture them from the person who is making the decisions, challenge them architecturally, and finally, present them in a model.

The CEO told you that the target outcome is to increase annual sales, and this can be achieved if *ABC Trading* is able to sell goods online directly to consumers instead of being only a retail sales company. Based on the answers that you have collected, you came back with the following diagram to have it approved by the CEO:

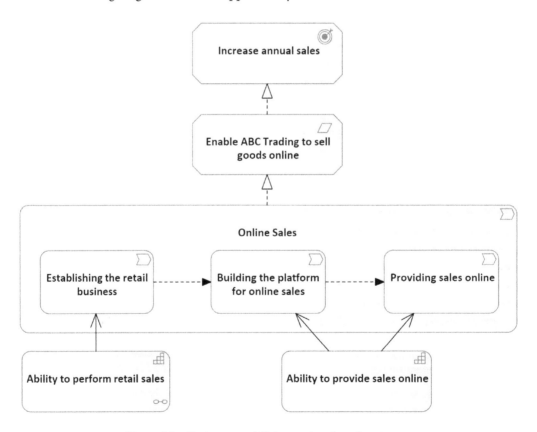

Figure 9.9 – Business capabilities serving the value stream

The diagram shows that offering online sales from a retail-oriented company requires performing three main steps (for the value stream):

- **Establishing the retail business**
- **Building the platform for online sales**
- **Providing sales online**

The online sales value stream will realize the requirement to enable *ABC Trading* to sell goods online, which in turn will realize the increase in annual sales outcome. The diagram also tells us that two capabilities need to be developed to deliver the value stream – the first is **Ability to perform retail sales**, and the second is **Ability to provide sales online**.

Important Note

Do not expect to get your diagrams right the first time. Final acceptance can take several rounds of refinements, so don't feel frustrated.

We have identified at a high level what we are planning to do and outlined how it is time to present much deeper details about the two required capabilities. We need to identify how to build and develop them and, ultimately, serve the value stream. In other words, we need to answer two questions:

- *What behavioral elements are needed to realize the identified capabilities?*
- *What structural elements are needed to realize the resources that will be assigned to the capabilities?*

Whether we have these elements today or not is another question to be answered later, but first, we need to identify what the needed elements are; therefore, let's start with the behavioral elements.

Realizing capabilities by behavioral elements

A capability is a behavior at the strategy layer, and it needs to be realized by one or more core behavioral elements from the business, application, and technology layers. If we want our business to have the ability to perform retail sales, it means that we want it to provide a specific type of services.

Without these services, a retail business will not be considered a retail business. The following diagram shows an example of the business services that realize the **Ability to Perform Retail Sales** capability:

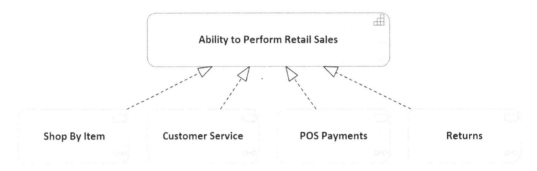

Figure 9.10 – Business services realizing a capability

Each of these business services can be realized by one or more behavioral elements, such as business processes, business functions, application services, and technology services. To realize the customer service business service, for example, you will need a set of functions and processes to realize it.

If it is automated, this means that it is realized by an application service or a technology service. In the following diagram, there is an example of a business service and how it can be realized by other business behavioral elements. The diagram also shows that the business service is completely realized by the **Customer Relationship Management** application service:

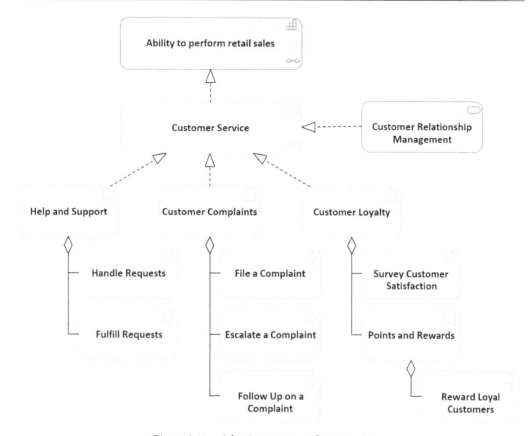

Figure 9.11 – A business service decomposition

If the business service is partially automated, it means that it is not realized completely by an application service. In this case, you must show the realization relationship only on the automated elements.

For every business service that you have identified, you can create a similar decomposition model, showing the elements that are composed (or aggregated). If your diagram grew in size and became busy with too many elements, you can put each main element in its own diagram.

After identifying the capabilities that we need to build and the behavioral elements that will realize them, it is time to identify which of these elements exist already and which ones don't.

Differentiating between the as-is and the to-be architectures

Whenever you plan for something, you will always have two architectural states at least – the present state and the planned future state. They are also known as the **as-is** and **to-be** states of the architecture.

We say at least two states because, in many cases, you will need some transitional states between the as-is and to-be architectures. One main reason for having transitional states is when the gap between the as-is and to-be is too large and requires building smaller steps at a time. One good example is when you are migrating from a legacy application to a modern one. You may build or use some components that are required only during the migration, but once it is completed, there will be no need for these transitional components.

One way to differentiate between the states in Sparx is to use the version value of each element. By default, every element you create in Sparx will be set to version **1.0**. Let's assume that the **Customer Relationship Management** application service does not exist now, and it is a future service in the to-be architecture if we want to fully automate the customer service business service.

To change the version to **2.0**, right-click on the **Customer Relationship Management** application service element and enter 2 . 0 in the **Properties** > **Properties** > **General** > **Version** field, as shown in the following screenshot:

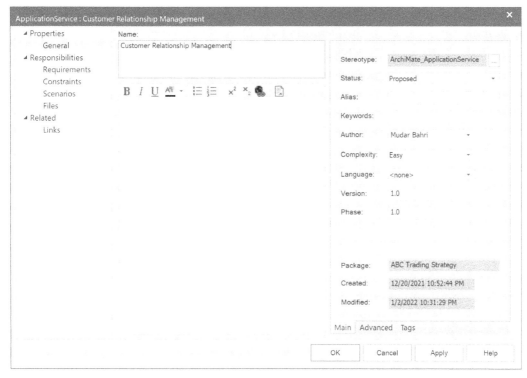

Figure 9.12 – Setting the version of an element

The **Version** field is a free text value, which means that you can put any value you want to use for versioning and not only numeric values. It is all up to you and the architecture team that you work with. You can use the values as-is, to-be, and in transition if you want. We will keep the default of 1.0 to indicate an as-is element, use 2.0 for the to-be, and 1.5 for any transitional element.

Do the same steps to change the version for the **Customer Loyalty** business function and all the composing elements. This indicates that this business function does not exist today as part of *ABC Trading's* internal functions, so it must be built and developed as part of the to-be architecture.

Now, we need to use color codes to visually differentiate elements based on their version. One way is to use the filters and layers as indicated in the *Diagram filters and layers* subsection in *Chapter 6*, *Modeling in the Technology Layer*. This is very useful if the diagrams will be accessed from Sparx. However, if you publish the image, the layers will be ignored and not published. So, we need a different way to color-code the elements based on their versions that are still effective even on printed diagrams and exported images, and this way is by using the **Legend** element.

From the **Common** section in **Toolbox**, find the **Diagram Legend** element and place it on the diagram. Double-click on the **Diagram Legend** element to open the dialog that you can see in the following screenshot:

Figure 9.13 – The diagram legend dialog

Follow these steps to create a color-coded legend:

1. Keep the default **Legend** value in the **Name** field or rename it if needed. This will act as the label that will be displayed on the legend box.

2. Make sure that the **Apply auto color** checkbox is checked. This tells Sparx that you want to apply special formatting on elements that match the criteria that we are going to define now.

3. From the **Filter** field, click the three dots and select **Element** > **Version**. This tells Sparx that your legend formatting criteria will be based on the version field of the diagram elements.

4. Leave the **Applies to** value as the default value, **<All>**. You can use this field to filter on a specific element type.

   Now, we need to tell Sparx to use three different color codes to differentiate elements based on the previously defined criteria. We will use **1.0** for **As Is**, **1.5** for **In Transition**, and **2.0** for **To Be**.

5. Enter 1.0 in the **Value** field and As Is in the **Display Value** field.

6. Select the **Fill Color** for version **1.0** elements, which I kept as the default white fill color.

7. Make sure that the **Apply Fill** checkbox is checked.

8. Since we áre not applying any color codes on the line (the element's borders) in this example, uncheck the **Apply Line** checkbox. We're also not applying any formatting on the borders' width, so we can uncheck the **Apply width** checkbox.

9. Click the **Save** button to save this legend value.

10. Click the **New** button and repeat *step 5* to *step 9* for the **1.5** and **2.0** versions. This time, select different fill colors for the two versions.

11. Click **Ok** to accept and close the dialog.

If you have done everything correctly, you can see how Sparx now differentiates the fill colors of the elements based on their versions, as shown in the following diagram:

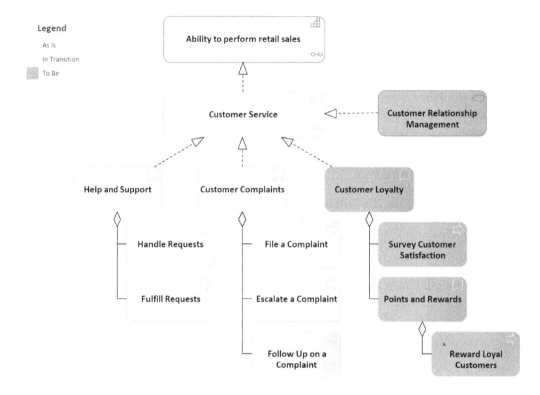

Figure 9.14 – A color-coded business service diagram

The legend element is reusable and can be placed on any other diagram in the repository. Furthermore, it will apply the defined color code automatically on the elements of the diagram that it is placed on. Try it yourself:

1. Select the legend element on the diagram.
2. Press *Alt + G* to find the legend element in the browser.
3. Drag the element and place it on any diagram that you have created earlier.
4. Change the version values of elements on the diagram.

You will see how the defined color codes will be applied automatically to the elements. Being a reusable element, changing values in the legend such as adding a new version or changing the colors will automatically be reflected in all the diagrams that use the same legend element.

The next topic will show us how to define the needed resources and assign them to the previously identified capabilities.

Identifying resources

Earlier, in the *Realizing capabilities by behavioral elements* section, we identified the capabilities that are required to serve the targeted value stream and the core behavioral elements that will realize the capabilities. In this section, we need to identify the resources that will be assigned to the identified capabilities and their realizing behavioral elements.

Since we went into more detail about the ability to perform the retail sales capability, we will continue using it for this example. You will still need to do the same for all the capabilities that you have identified. For *ABC Trading* to be able to perform retail sales, it needs to have the following types of resources:

- **Business actors and roles**: The managers, accountants, security guards, janitors, forklift operators, salespersons, and all the other required human resources to enable this capability. You can nest business actors and assign them to business roles.

- **Information**: No matter whether the information is documented on paper, in a knowledge base system, or undocumented, with the proper knowledge, people will be able to deliver value. Information about supply and demand, trending products, market changes, team-building techniques, project management, operations, and other information is essential to be capable of performing retail sales.

- **Interfaces**: Customers will need interfaces to be able to use services. These interfaces can take the form of a showroom, a call center, a web interface, and many other forms of human and application interfaces. Remember that when a business interface is realized by an application or a technology interface, it indicates that it is an automated (or digital) interface.

- **Technologies**: These can be anything from heavy machinery, security doors, badge scanners, and ceiling fans to computers, network devices, and computer applications – the list is long.

The following diagram shows sample resources assigned to the ability to perform the retail sales capability. We can now identify which resources we have and which resources we need to have in the future (to-be). We will identify the ones in the to-be architecture state by changing their version value to 2.0, and the ones in transition by the 1.5 value as a version:

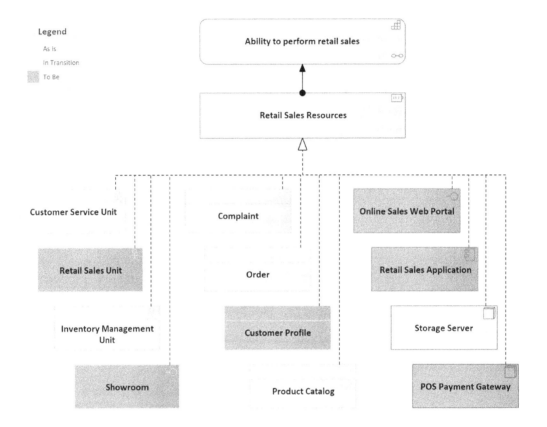

Figure 9.15 – Retail sales resources assigned to a capability

Note that we have reused the same legend element that we created in *Figure 9.14*, so it automatically applied the fill colors on elements based on their version.

To recap, enterprise resources are realized by structural elements from every enterprise layer (business, application, and technology). Capabilities are realized by behavioral elements from every enterprise layer. Resources are assigned to capabilities to realize and influence requirements. Capabilities with the resources assigned to them realize the courses of actions, which will help the enterprise to achieve the desired outcomes.

The differences between the two versions of the architecture mean that there are gaps between the two. To transform from an as-is state to a to-be state, these gaps must be closed. In the next section, we will see how to identify gaps, model them, and plan on closing them.

Introducing implementation elements

Without being implemented, all your architectural work and artifacts will remain theoretical. Good presentations and nicely published enterprise content are great achievements, but they will not make things happen. Your next action is to convert your architectural artifacts into actionable plans and start building what you have been planning for. In this subsection, we will introduce you to the implementation elements and how to use them to model your plans, so let's start with the plateau element.

Defining plateaus

"A plateau represents a relatively stable state of the architecture that exists during a limited period of time" (https://pubs.opengroup.org/architecture/archimate3-doc/chap13.html#_Toc10045450).

A plateau represents the state of the architecture. It can be a past, present, or future state. When we mentioned the as-is, in transition, and to-be architectures in the previous *Introducing strategy elements* section, each one of these three states can be represented with a plateau element. You can also think of plateaus as phases or milestones within a bigger effort to achieve goals and address requirements. Let's look first at the plateau's focused metamodel before elaborating:

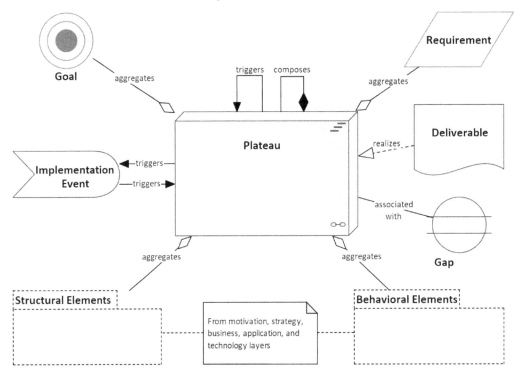

Figure 9.16 – The Plateau focused metamodel

As you can see in the focused metamodel, the plateau can aggregate any number of structural and behavioral elements from any layer of the enterprise. The plateau can compose a bigger plateau, and it can be composed of many smaller plateaus as well. It is always recommended to put the right amount of information in a single diagram. If you look back at *Figure 9.9*, which models the value stream, you can consider modeling a plateau for each value stream element, as shown in the following diagram:

Wholesales Only   Wholesales and   Wholesales, Retail,
                              Retail   and Online

Figure 9.17 – Three plateaus representing high-level transformation phases

Within each plateau, you can identify what elements will belong to it. The same elements can belong to any number of plateaus. Transitioning from an as-is plateau to a to-be can involve keeping all or some of the current elements to the targeted state. In real-life examples, a plateau can contain hundreds of elements, which makes it difficult to fit them all in one diagram. It is highly advisable to make diagrams that convey a single idea at a time. The following diagram shows two plateaus with some shared and new elements:

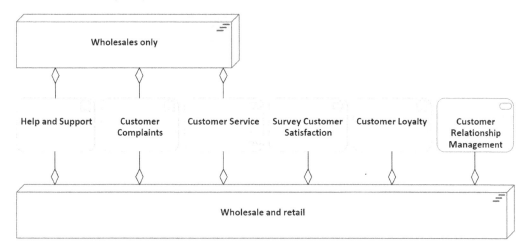

Figure 9.18 – The difference between two plateaus

As you can see, there are elements that are part of the as-is Wholesales Only plateau but that will continue to be used in the targeted Wholesale and Retail plateau. There are also elements that belong only to the targeted plateau, and those represent the gaps that must be closed. If there are elements in an as-is plateau but not in the to-be, they indicate elements that will be retired or decommissioned from the enterprise. Let's talk more about gaps in the next subsection.

Defining gaps

"A gap represents a statement of difference between two plateaus" (`https://pubs.opengroup.org/architecture/archimate3-doc/chap13.html#_Toc10045451`).

Gaps are passive structure elements, so they are a type of data or information. This data concerns the decision-makers, as they can prioritize the gaps and allocate the required resources to them. If you have one plateau representing the as-is and another plateau representing the to-be, the differences between the two are described by the gaps that need to be closed to complete the transition. Gaps usually indicate missing elements. However, they can also indicate additional elements that will be retired from the as-is element and will not be taken forward.

The following diagram shows the focused metamodel of the **Gap** element:

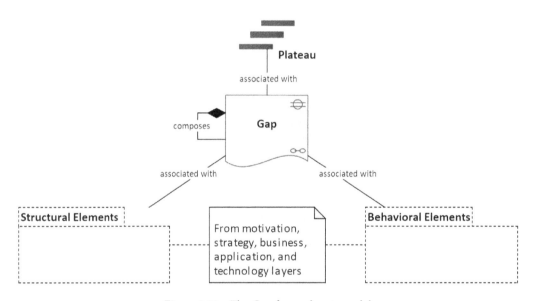

Figure 9.19 – The Gap focused metamodel

An example of how gaps between two plateaus can be modeled is depicted in the following diagram:

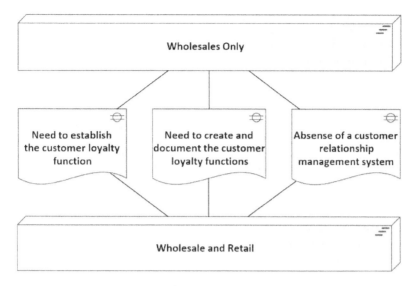

Figure 9.20 – Modeling gaps between two plateaus

Gaps can be associated with any structural and behavioral elements from any enterprise layer. You can have a single gap that involves some missing roles, processes, strategies, applications, and technologies, of any number, as shown in the following diagram:

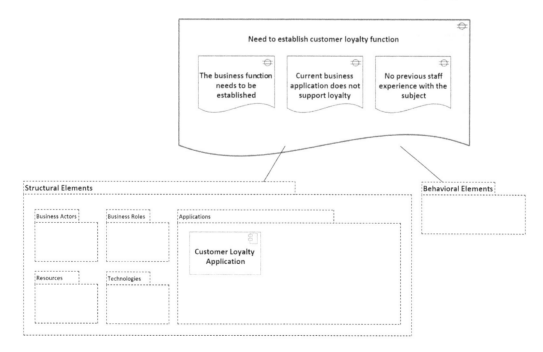

Figure 9.21 – Modeling nested gaps

As you can see in the diagram, we have broken down one of the gaps into smaller gaps to ease turning them into smaller actionable items known as **work packages**. The diagram also indicates that closing this gap will use the resources that are allocated for building the ability to perform retail sales. You can develop as many diagrams as needed where each can highlight a specific point of interest. Some diagrams will show how many business roles the enterprise needs to be capable of to close the gap. Another diagram can show us what processes we need to learn, develop, and automate. A third diagram can show what technologies are missing and are required.

At any point in time, you may have limited information, so you add what you know and keep adding elements whenever you learn something new. Developing a new capability is a learning experience for everyone. If time is not on your side, then you have to use part of the allocated resources and hire people to fast-track the process.

We briefly mentioned work packages, but we didn't talk in detail about them, so let's do that now.

Defining work packages

"A work package represents a series of actions identified and designed to achieve specific results within specified time and resource constraints" (https://pubs.opengroup.org/architecture/archimate3-doc/chap13.html#_Toc10045447).

Work packages are internal behavioral elements. If you are familiar with the **Project Management Body of Knowledge** (**PMBOK**), you can think of work packages as the equivalent of work breakdown structure items. From this point in the architecture definition, it will be very helpful to involve the project managers to participate in defining and modeling the work packages and their composition hierarchy. Work packages can be composed of smaller ones, and there is no limit to how small a work package can be. Here is the **Work Package** focused metamodel diagram:

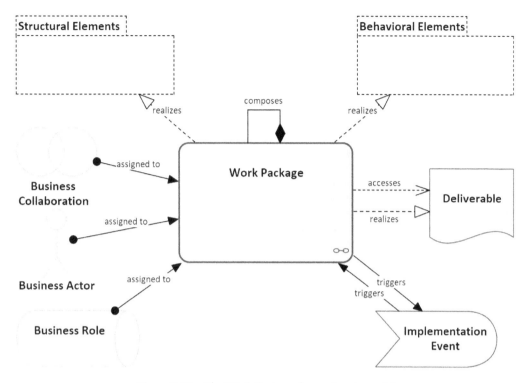

Figure 9.22 – The Work Package focused metamodel

You need to avoid making them as granular as the tasks are, so they still need to maintain the concept of *packaging* multiple actions that together will realize a deliverable.

We need to remember that Sparx is not a project management software, so we don't need to move all project management activities into it. Even with the ability to create some models such as Gantt charts, Sparx is still an EA repository, and artifacts must remain as EA artifacts. Once a project is kicked off, the project manager can use more sophisticated project management software to manage the project while maintaining the trace back to the related EA artifacts, such as the work packages and the deliverables.

The following example explains how a work package can be composed of many smaller work packages, each one realizing some of the elements that are associated with a gap:

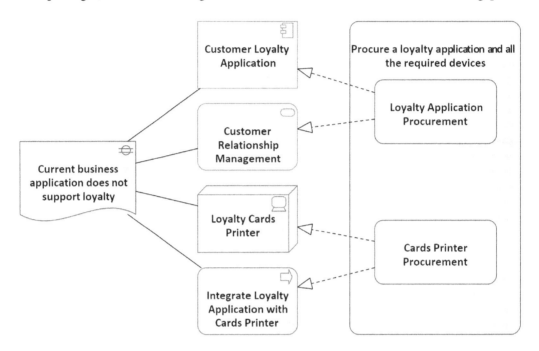

Figure 9.23 – Work packages realizing gap elements

The next element is the deliverable, which is the result of completing a work package, so let's learn more about it.

Defining deliverables

"A deliverable represents a precisely-defined result of a work package" (https://pubs.opengroup.org/architecture/archimate3-doc/chap13.html#_Toc10045448).

Deliverables are passive structure elements, so they are treated the same as data or a piece of information. They can be anything that a work package is intended to deliver. This means that they realize either structural or behavioral elements based on what work packages were intended to deliver. Deliverables can be tangible, such as products, software, modules, documents, and reports. They can also be intangible, such as the maturity of an organization, customers' satisfaction, or employees' loyalty to their workplace.

Deliverables are like milestones that, when achieved, mark the end of a work package. Based on the size of the work package, a deliverable can mean the completion of a single user story, a whole project phase, or the entire project. Remember that these requirements were originally aggregated within a plateau, so completing the deliverables means that we have realized all parts or a part of the plateau. It's all based on the requirements that the deliverable has realized.

The following diagram contains the focused metamodel of the deliverable element:

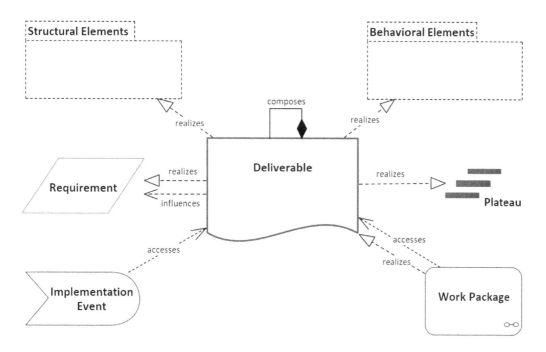

Figure 9.24 – The deliverable-focused metamodel

Look at the following example to understand how a work package realizes a deliverable, which itself realizes a plateau:

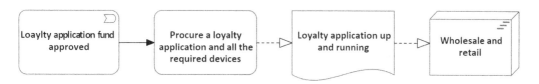

Figure 9.25 – Work packages realizing a deliverable, realizing a plateau

The left-most element is an implementation event, which we will talk about in the next subsection. It triggers the execution of a work package, which once completed will realize a deliverable, which in turn realizes a plateau.

The last element that we will discuss in this chapter is the implementation event before wrapping up with the summary.

Defining implementation events

"An implementation event represents a state change related to implementation or migration" (https://pubs.opengroup.org/architecture/archimate3-doc/chap13. html#_Toc10045449).

Implementation events are events that initiate a change in relatively stable plateaus or work packages. They are instantaneous and have no duration, but they may initiate changes that last for long periods of time. The following diagram shows the implementation event-focused metamodel:

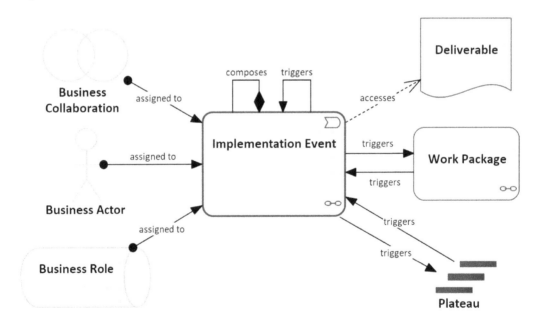

Figure 9.26 – The Implementation Event focused metamodel

Internal business structural elements are assigned to implementation events to manage and control their flow, as shown in the following diagram. This could be a business unit, a specific person, a specific role, or a combination of all:

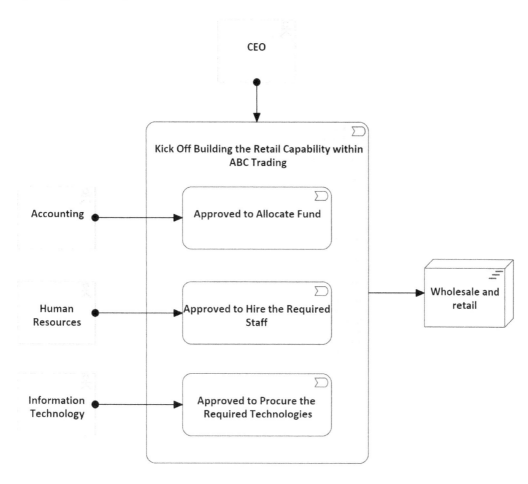

Figure 9.27 – Business structural elements assigned to implementation events

With this, we have introduced the last implementation elements, and we have reached the end of this chapter as well. Let's summarize what we have achieved.

Summary

Strategic decisions are usually made at the highest level of an organization. Enterprise architects can support decision makers best by helping them to visualize their thoughts and ideas by turning them into diagrams. It can take you multiple rounds to put what's in their heads on paper, but this effort will be appreciated if done right.

Implementation plans including phases and deliverables can also be modeled. Models should not include all the details that project management software provides, but they need to remain at the enterprise architecture level. The implementation architecture artifacts should align with other enterprise elements, while project management software will help in daily project management activities. These are two different worlds that need only to keep the traceability between them.

By reaching the end of this chapter, you are capable of building an EA repository using Sparx, or maybe you have already done that while practicing with the provided examples.

In the next chapter, you will learn how to keep this repository clean and up to date. It took you a lot of effort to build it, and you don't want to waste it.

Section 3: Managing the Repository

This section will show you how to capitalize on the content that you have developed in the repository and guide you on how to keep it up to date and share it with a larger audience.

Building an enterprise architecture repository is a dynamic process that usually starts small and builds up as you progress, so making changes to its structure is a very common occurrence. You need to know how to properly adopt these changes without breaking the integrity of the repository and risking losing your spent efforts.

Additionally, you have to keep in mind that most of your audience are not experts in using Sparx Systems Enterprise Architect, so being able to share your work with them using documents and web content will help in delivering the value of your enterprise architecture to a larger audience.

This section comprises the following chapters:

- *Chapter 10, Operating the EA Repository*
- *Chapter 11, Publishing Model Content*

10
Operating the EA Repository

Someone once said that imitation is the sincerest form of flattery. Once you begin producing and publishing models based on detailed, complete, and consistent information about your organization's enterprise, there is little doubt others will want to do the same. This is an ideal situation. Having a common and consistent understanding of the composition and state of your enterprise is the basis for making sound decisions regarding it. It is what brings trust to the **Enterprise Architecture** (**EA**) practice. Indeed, you should encourage reusing the information in your repository whenever possible.

Sharing a repository, however, presents another set of challenges. For anyone who has worked within or managed a development environment or any other shared resource, these challenges may seem familiar. Managing or operating an EA as a shared resource is the subject of this chapter. In this chapter, we will review three topics related to operating the EA repository. Those topics are as follows:

- Sharing repositories
- Managing a shared repository
- Repository governance

Much of what we will cover in this chapter is covered in detail in the Sparx User Guide. However, we've simplified some of the most important features here in this chapter in a condensed format to make you aware of them and provide references to more information, and we've also added our own bits of wisdom to these various practices. Using a shared repository facilitates many interesting features of Sparx such as discussions, reviews, chat, mail, and the calendar. We won't go into detail about these features in this chapter. As you will see, their use is very straightforward, and they don't need much explanation. Instead, we've focused on the issues we think are critical to successfully deploying and managing a shared repository.

Let's take a look at what you will need to complete this chapter.

Technical requirements

We hope by now that you have **Sparx Systems' Enterprise Architect** installed on your computer.

You can access the repository at the following GitHub address: `https://github.com/PacktPublishing/Practical-Model-Driven-Enterprise-Architecture/blob/main/Chapter10/EA%20Repository.eapx`.

This chapter discusses repositories deployed on a commercially available relational database engine such as **Oracle**, **SQL Server**, or **MySQL**. If you have no intention of doing so, you can skip this chapter. If you do plan to migrate to a shared repository, you will need to read the Sparx User Guide covering this topic. You can find it at the following location: `https://sparxsystems.com/enterprise_architect_user_guide/15.2/model_repositories/settingupdatabasemodelfile.html`.

> **Important Note**
> You will need administrator privileges on your repository to take full advantage of this chapter.

Sharing repositories

The most effective way to share repository access is to maintain your repository in a form that is easily shared – a **relational database**. Even if your repository is an `.eap` file on your desktop, it's already in a relational database. Sparx uses a JET database to store local repositories. While that's fine for local access, sharing repository access among five or more users is best accomplished using Oracle, Microsoft SQL Server, MySQL, or a similar database engine to house your repository on a shared remote server.

Sparx provides two ways to connect to such a remote repository:

- Using a native database connection method
- Via an HTTP server called **Pro Cloud Server**

There are benefits and drawbacks to either choice:

- Connecting via the direct database connection is the fastest, but it requires a database user account for each user. It also requires that each user has network access to the server on which the database engine is deployed. This may not always be the case.
- Pro Cloud Server acts as a proxy and removes the need for a direct connection to the database engine. Communication with Pro Cloud Server is via HTTP, which is ubiquitous.
- The downside is that Pro Cloud Server adds another layer of administration and a very small amount of performance overhead to translate from HTTP.

Deciding which way to connect to your repository is beyond the scope of this chapter. It suffices to say that you need your enterprise repository stored in a relational database that you can share. Whichever method you choose, you must also consider how you will support multiple parallel tracks of model development.

Organizing and reorganizing the repository

Sharing and reusing model information requires that the modeler be able to find that information. The structure of your repository is critical to finding information. As your repository grows, its structure becomes even more important. Without a proper structure, the repository will become a tangled mess that is of little use to anyone. There is no single layout that is appropriate for all repositories, nor is one structure appropriate for any given repository over its entire lifespan. Analyzing, assessing, and reorganizing a repository needs to happen often. You should create a process for reorganizing as part of your EA practice.

As you begin the process of modeling, whether in a new repository or on a new subject area, it's often fine to simply place all elements and diagrams for the model into a single package. As you add models and subject areas to your model, you need to consider how you will reference your new models or subjects. When the number of items in a package begins to make it difficult to find things, it's time to reorganize.

The first thing to consider in organizing your repository is what areas modelers are most likely to reuse. They are more likely to reuse elements than diagrams. Diagrams represent a view of a subject from a specific viewpoint. The model element represents the facts that are often immutable. Separating model elements from the views that use those elements is usually the first level of organization.

Once the number of elements in a package starts to become unwieldy, you should consider organizing the elements by element types. You can use the ArchiMate® stereotypes for this purpose. This has the added advantage of allowing you to easily perform enterprise-wide operations on the repository based on the element type. We saw a glimpse of this type of organization back in *Chapter 7, Enterprise-Level Technology Architecture Models*, when we organized equipment elements by type:

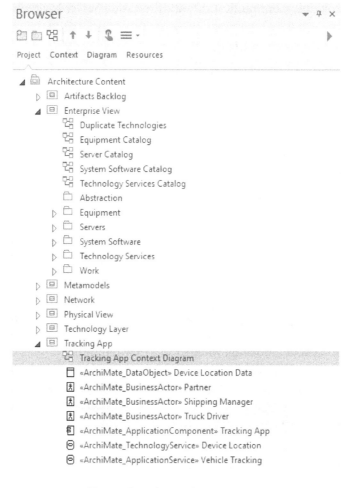

Figure 10.1 – A repository structure

With all of your elements organized by element type, you may be wondering what happens when you need to reference these same elements in other ways, such as from the context of an application or a business process. This is where a diagram comes into play. A diagram that contains elements from several packages serves as an index to those elements. Sparx even provides a way to open the diagram as a list. From any open diagram, right-click inside it and select **Switch View**, and then select **Switch to List View**, as shown in the following figure:

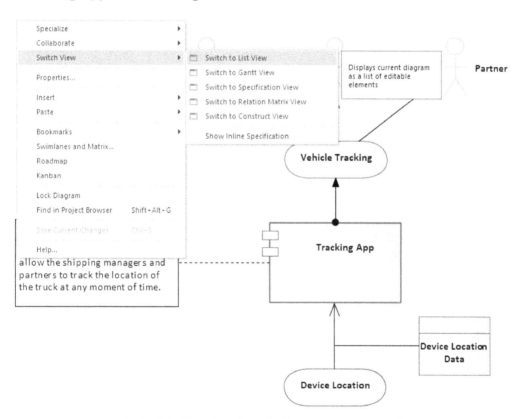

Figure 10.2 – Switching views from the Tracking App Context diagram

The result is a neat list of elements, as shown in the following diagram:

	Name	Status	Type	Modified
⬭	Vehicle Tracking	Proposed	Activity	6/5/2021
▤	Tracking App	Proposed	Component	5/11/2021
▦	Truck Driver	Proposed	Class	5/11/2021
▦	Partner	Proposed	Class	5/11/2021
▤		Proposed	Note	5/14/2021
▦	Shipping Manager	Proposed	Class	5/11/2021
▤		Proposed	Text	5/17/2021
⬭	Device Location	Proposed	Activity	6/5/2021
▦	Device Location Data	Proposed	Class	6/5/2021

Figure 10.3 – A list view

While this approach to structuring a repository has served us well, it's not the only way. Our intent is not to be prescriptive in the structure you choose but rather to encourage you to apply some time and thought to the subject of repository organization. Now, let's look at other important considerations for sharing repositories.

Model abstraction

As you model various aspects of your enterprise, the number of connections among elements increases dramatically. This is especially the case when you increase the number of modelers working concurrently. These connections can often become troublesome. Consider the following scenario.

One of your architects, who we'll call *Sam*, is in the process of modeling a new inventory management system called ATC-INSTOCK. You, on the other hand, are responsible for identifying critical changes to the enterprise accounting system, ATC-GL. The ATC-GL model is comprised of 17 different elements, representing the components that make up the accounting system. One of these elements is the ATC-API1 component, which provides the API for external systems to add ledger transactions. *Sam* models the interaction for adding inventory by linking ATC-INSTOCK to ATC-API1. You realize that you need to replace the ATC-API1 component with a new type of element. Because ATC-GL has been around for years, there are dozens or even hundreds of links to ATC-API1. How do you reconcile these links with the latest version of ATC-GL?

There are issues with this scenario that would justify objection, not the least of which is the question of why two enterprise architects are working at such a low level. However, that's a question for another book; the question for this book is why there is no abstract element to represent the entire `ATC-GL` system. This is the point of this subsection. There need to be abstract representations in the model for all major information systems. *Sam* needs to reference the enterprise accounting system, not the `ATC-API1` component. You can change all of the components or elements that make up the accounting system, but you should never remove or delete the abstract element that represents the entire enterprise accounting system.

Model replication

One of the primary tenets of a good data management or data architecture strategy is to eliminate redundant data whenever possible, except for backups. This is no different for model data. Replicating models is not a good practice. If you replicate a model to allow parallel model development, inevitably, both models will become inaccurate or incomplete over time. Reconciling the differences between replicated models can become a huge task. Our recommendation is to not go there. One exception to this rule involves the use of a version management tool. We'll talk about that next.

Version management

Sparx interfaces with various version management systems to manage parallel model development. This feature can be convenient in allowing for offline model work. It does, however, require effort when bringing two parallel development streams back together. As with any source management tool, when checking-in offline work, conflicts can occur.

You must reconcile those conflicts manually before merging the two branches. This process is similar to merging two branches of source code using a version management system. The only difference is that the content you are merging with may not be as familiar to you as the source code is unless you've worked with Sparx for a while.

Sandboxes

One method of supporting parallel model development that has worked quite well for us in the past is the employment of what we call sandboxes. A **sandbox** is simply another model root in the same repository, dedicated to a project that will change the enterprise. We name these sandboxes according to the project they represent. The following diagram depicts a new sandbox to support the new inventory management system at *ABC Trading*:

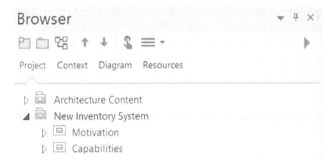

Figure 10.4 – A new sandbox for the inventory system

Naming a Sandbox

Although the name *sandbox* implies a play area, we avoid naming these new model roots after a person, as we've found that it tends to encourage experimentation in the shared repository. It's best to leave experimentation to a local repository; otherwise, these *sandboxes* tend to turn into *cat boxes*, if you know what we mean.

The use of sandboxes in the shared repository provides us with the means to represent the potential or proposed future state of the enterprise while leaving our original model root to represent the current state of the enterprise. This is an important capability. We need to always maintain the ability to know and report on the current state of the enterprise. Keeping new or proposed changes in a separate area is one measure to ensure that you always have that ability. Now, we'll look at another necessary measure – maintaining status.

Element status

One of the advantages of using a sandbox in a shared repository is that we can still reference the existing architectural elements. Sparx allows us to establish links from elements in one model root to elements in another. To do this safely, however, we must employ a means of identifying our sandbox elements as *new*; otherwise, we could run into a situation where reports run against the existing model root may unintentionally include new elements from a sandbox. We will learn more about reports and report templates in the next chapter. For now, we just need the means to distinguish elements other than by their location. This calls for the use of the element **Status** field.

When you create a new element in Sparx, the default behavior sets the element's status to **Proposed**. This is the perfect status for working in a sandbox. Sparx comes preconfigured with a set of status values. To see the configured status values, navigate to **Configure** > **Reference Data** > **Model Types** > **General Types** and select **Status** from the left panel of the dialog, as follows:

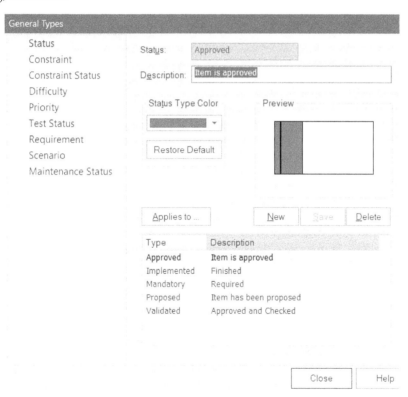

Figure 10.5 – The element status value configuration

To distinguish those elements that represent the current state of the enterprise, we set their status to **Implemented**. Before you go and open the properties dialog for each element in your repository, Sparx provides a quick and convenient way to update the status for an entire package, or even an entire model. Simply select the package level for the elements you want to update, and then navigate to **Design** > **Model** > **Manage** > **Package** > **Update Status**. The following dialog will appear:

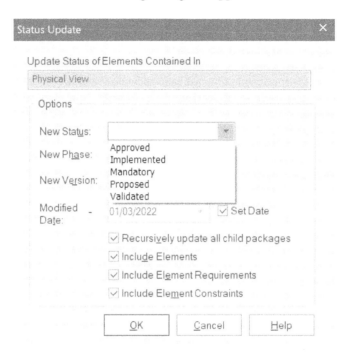

Figure 10.6 – The Status Update dialog

Select the status you want to change to and click **OK**. As you can see, this dialog can change the values for the **New Status**, **New Phase**, **New Version**, and **Modified Date** fields.

The use of the element **Status** field implies a process behind the creation of elements and their promotion from a sandbox into the core model. Before you begin using the **Status** field, it's best to define and document the process of promoting elements from the sandbox. One way to do this is through the use of a UML state machine diagram, as depicted in the following figure:

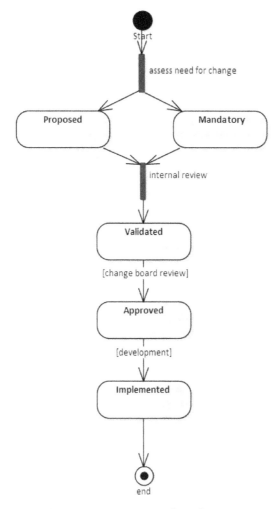

Figure 10.7 – A state machine diagram

A **state machine** diagram describes how an entity, such as an element, responds to input, processes, the current state, and events that occur by showing the resulting state of that entity. In our example, state is analogous to status. We can show how the status of an element changes based on its current status and what process has occurred. The links between statuses are the processes or activities that take place to change the element from one status to another.

We'll talk more about architecture-related processes in the section called *Repository governance* later in this chapter.

Managing the shared repository

As the use of the shared repository grows, so grows the need to manage and care for it. Almost immediately on implementation, the shared repository needs an administrator or manager. In the beginning, we've usually relied on one or more knowledgeable users to perform administration and configuration tasks. As the use and number of users grow, you may find that this needs to change. In this section, we will take a look at the most common administrator tasks and functions. The topics we will cover in this section include the following:

- Configuration management

- Implementing security

- Deleting and merging elements

- Backup and restore

- Automation

Configuration management

Many of the Sparx administration tasks can be found in the **Configure** tab, as depicted in the following screenshot:

Figure 10.8 – The Sparx Configure tab

As you can see, the functions on this tab relate to enterprise-wide settings or the health of the repository. From here, you can configure security, version control, status values, authors, data sources, and other similar functions.

Often, the administrator is responsible for creating report templates and publishing reports and HTML content. These functions can be restricted to a Sparx user with administrative privileges only. We'll look at implementing security next.

Implementing security

The default configuration of a Sparx repository has security turned off. You must be a registered user to turn security on. You won't be able to do it by using a trial version of Sparx. To turn security on, you must obtain an access key from the Sparx Registered User web pages at `https://sparxsystems.com/registered/reg_ea_down.html#RoleSecurity`. You then enter this access key into the dialog that appears when you select the **Enable Security** option from **Configure** > **Security** > **Administer** > **Enable Security**:

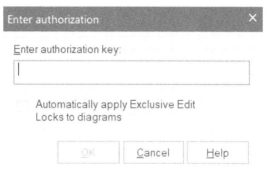

Figure 10.9 – The authorization key entry dialog

Sparx provides the ability to set up permission for a single user or a group. Setting up groups allows you to specify permissions at the group level and then assign individuals to the appropriate group. You can import users into a new group from a **Windows Active Directory group**. This works for the initial setup of the group only. After that, you'll need to add each user manually. You can also implement a single sign-on from the user permissions configuration dialog by selecting either **Accept Windows Authentication** or **Accept OpenID Authentication**. The following screenshot depicts the **Security Users** permissions dialog:

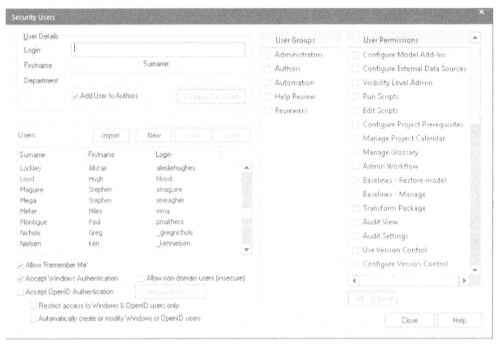

Figure 10.10 – The Security Users permissions configuration dialog (image courtesy of the Sparx User Guide:)

To learn more about enabling security, refer to the Sparx User Guide, which can be found at `https://sparxsystems.com/enterprise_architect_user_guide/15.2/modeling/usersecurity2.html`.

Model locks

If a user has the **Lock Elements** permission granted, they can lock or unlock elements, diagrams, or packages from getting updated by other users. It's also possible to set group-level locks that prevent updates from anyone outside of that group. To set a lock, select the element, diagram, or package in the project browser, and then right-click and select **Lock** from the context menu.

The browser window indicates when the current user has established a lock on an element, package, or diagram by placing a blue-colored exclamation point next to the locked element. It identifies locks established by other users with a *red exclamation point*, as shown in the following screenshot:

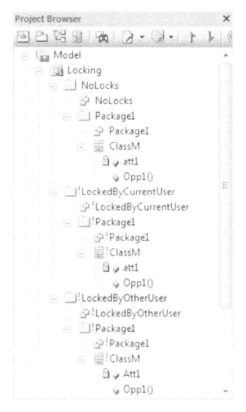

Figure 10.11 – The browser window lock indicators (image courtesy of the Sparx User Guide)

On occasion, locks can become stranded. When this occurs, the administrator can clear those locks by accessing **Configure** > **Security** > **Locks**.

Deleting and merging elements

Deleting elements from the repository can be a risky activity. An element can have dozens or even hundreds of links to other elements in the repository. The element can appear on any number of diagrams. These links and diagrams are not readily apparent unless you look for them. There is no *undo* function in Sparx. When you delete an element from the project browser, it's gone.

The only way to get it back is by restoring it from a backup or recreating it. For these reasons, you need to take extra care when deleting elements. In this section, we will show you three ways to check for links and whether the element appears in any diagrams. The method you choose will often depend on how many elements you need to delete.

When deleting a single element, the following steps describe the simplest method to check for usage in a diagram:

1. In the **Project** browser, select the element and right-click on it:

Figure 10.12 – Right-clicking on an element

2. After that, from the right-click menu, select **Find in all Diagrams…**:

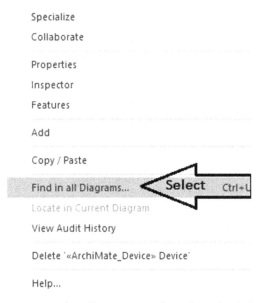

Figure 10.13 – Find in all Diagrams… from the right-click menu

The following window will open after we click on **Find in all Diagrams…**:

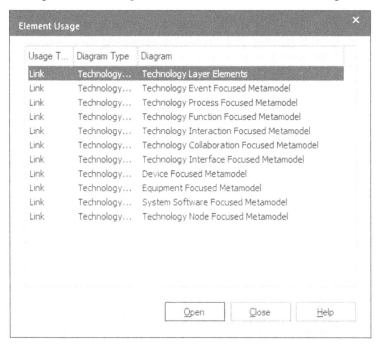

Figure 10.14 – Element Usage

The prior example indicates that the element in question appears in 11 diagrams. From this dialog, you can open each diagram. To check for links, open the element's **Properties** dialog and select the **Links** option from the leftmost panel. As the following example shows, the element in question links to 14 other elements:

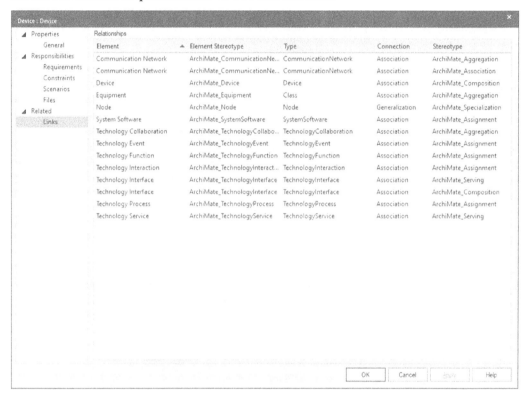

Figure 10.15 – The element links in the Properties dialog

When you need to delete several elements, the following process might be a bit easier:

1. Select the first element in the **Project** browser.

2. Open the **Traceability** panel by selecting **Design** > **Impact** > **Trace**.

3. The **Traceability** panel displays all links to and from the selected element.

4. To see a list of diagrams that the element appears in, click on the **Usage** button, as shown in the following screenshot of the **Traceability** panel:

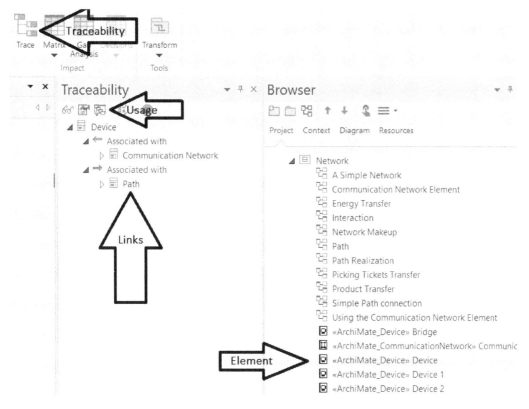

Figure 10.16 – The Traceability panel display method

5. Select the next element to research in the **Project** browser and repeat from *step 3*.

Merging elements

Sometimes, the need to delete an element arises because someone inadvertently created an extra version of an element. In such a case, it's necessary to move all the links from one of the extra elements, which we call the source, to the other, which we call the target. We also need to replace the source element with the target in any diagram. We call this process **merging**. The following steps describe the best way to accomplish this:

1. Create a new temporary work diagram.

2. Drag both source and target elements onto the diagram.

3. Select the source element on the diagram.

4. Right-click and select **Insert Related Elements…**.

5. Repeat *step 3* and *step 4* for the target element.

6. Select a link endpoint on the source element.

7. Drag the link to the target element.

8. Repeat *step 6* and *step 7* for all remaining links on the source element.

9. Check for usage of the source element. Replace all occurrences of the source element with the target element in any diagram.

10. When clear, delete the source element.

As you can see, this process is a bit more involved than simply deleting the element. This is necessary because there is a great deal of information packed into the elements of a model.

Of course, if you do accidentally delete an element, you can always restore that element from a backup. We'll cover that topic next.

Backup and restore

There are many advantages to using a shared repository, but there is just not enough time and space to cover all of them in this chapter. Here, we will focus on one advantage, which is the opportunity to use the backup processes that are already in place. If you're deploying to a relational database platform that your organization already uses for other systems, you likely have database administrators and backup cycles in place that can cover the Sparx repository. If so, you're in luck. If not, you will need to create them.

In addition to database backups, you can also use a feature of Sparx called **Controlled Packages** to export portions of a repository for storing offline or to a version control system. Sparx stores the exported information in XML format. For a comprehensive description of this feature, see the Sparx User Guide section for Controlled Packages at the following link: `https://sparxsystems.com/enterprise_architect_user_guide/15.2/model_publishing/controlledpackages2.html`.

Automation

Automation is what Sparx calls the facility for providing scripting access to its object model. You can use the automation interface to provide functionality that Sparx doesn't provide out of the box. We won't go into detail on automation here because the User Guide provides comprehensive coverage of this topic. You can find it at the following location: `https://sparxsystems.com/enterprise_architect_user_guide/15.2/automation/automation_interface.html`.

We can tell you about occasions where we've used the automation interface to solve specific problems. The automation interface is great for making changes to large numbers of elements. In one case, we needed to change the stereotype applied to hundreds of elements in a model. Without automation, we would have had to open the properties dialog for each element.

On another occasion, in a model that contained elements with long and complex chains of relationships, we needed to identify elements of one type that had indirect (12 elements deep) relationships to elements of another type. While we could do this for an individual element by navigating the **Traceability UI**, in this case, the stakeholder didn't have Sparx, and there were hundreds of these elements. In both cases, we used **JScript** to make the change and produce a report.

If you're interested in the internal structure of the Sparx repository, the publication *Inside Enterprise Architect: Querying EA's Database* by Thomas Kilian is available at `http://leanpub.com/InsideEA`. In it, he lists all of the Sparx internal tables that he's discovered and identifies their content and use.

Repository governance

Argh! It's the dreaded G-word. Yes, governance is upon us. What would a book on EA be without a reference to governance? We promise to be brief. Our primary intent is to make a distinction between architecture governance and repository governance.

Architecture governance

Architecture governance is about making architectural decisions. This subject usually includes such areas as the following:

- Governing boards
- Principles and policies around architecture
- Governance review processes and documentation

In-depth coverage of architecture governance is beyond the scope of this book. There are already tomes written on this subject. You can see what Sparx Systems says about governance at the following URL: `https://sparxsystems.com/enterprise_architect_user_guide/15.2/guidebooks/ea_managing_an_enterprise_architecture.html`.

While we have our own perspectives on governance, we would prefer to not muddle this book with yet another opinion. We hope you will appreciate that position. That said, we would like to express a couple of principles related to governance that we have found to be invaluable in our personal careers as enterprise architects.

The best information makes for the best decisions. Some organizations entrust us, as architects, with making the architectural decisions for their enterprise, but that's not always the case. The larger or more complex the environment, the more people are often involved in the decision-making process. Regardless of whether we or an architecture board make the decisions, we are always responsible for the information on which those decisions are based.

Our aim in authoring this book is to help you do the following:

- Create the best information possible regarding your enterprise.

- Create enterprise information using the easiest and most consistent means possible.

- Make comprehensive and accurate information readily available to your stakeholders and decision-makers so that they can make the best decisions possible.

- Garner trust in the processes and methods you use to create enterprise information by using open and standard-based information artifacts.

- Provide open, accurate, and complete information on architecture alternatives, some of which you may not always agree with.

Okay, that's it! We will now step down from our soapbox. Now, let's look at repository governance.

Repository governance

Repository governance is a less lofty subject than architecture governance. This is all about making decisions about the use and maintenance of the shared repository. As you acquire more Sparx users, repository governance becomes more important. In this case, the decision-makers are well known.

Those who use the repository should have a say in how it's used. If the number of architects using the shared repository is so large that they can't all be involved in the decision-making process, then you'll need to establish a decision-making board to do it. If that's the case, congratulations! You are well past the governance advice provided here in this book. More likely, you are just starting out with your shared repository. Here are the areas in which you may need to employ a small governance process:

- Repository structure or restructuring

- Configuration settings and changes

- Sandbox creation and maintenance

- Diagram and element styles, including which colors and fonts to use
- Standard report templates and template structures (see *Chapter 11*, *Publishing Model Content*)

This doesn't need to be formal. Quite often, this can be done as a discussion point after a regular weekly architecture meeting.

Summary

Using Sparx Systems' EA on a shared repository is an effective means of generating and maintaining information about your enterprise, both in its current state and any proposed future states. As with any information system, the shared repository needs regular care and attention.

Also, as with any other information system, the value of the information contained within a shared repository quickly becomes apparent as it matures. The repository deserves attention commensurate with that value. It is that care and attention that we have attempted to cover in this chapter. We hope you've found it helpful.

In the next chapter, we will cover the ways in which you can make this valuable information known to others.

11
Publishing Model Content

Once you've captured information in your **Sparx** repository, you will inevitably want to make that information available to your stakeholders. Expecting your stakeholders to have the ability or desire to navigate your repository using Sparx is not practical. They need information in a form with which they are comfortable. Fortunately, this is an area where Sparx, once again, shines, which brings us to the subject of this chapter – **publishing model content**.

In this chapter, we will learn different approaches to publishing model content for consumption by others in your enterprise. Unlike previous chapters, where we explored the use of the **ArchiMate®** language and the best ways to use it, this chapter is all about the publishing features of Sparx.

This chapter includes the following sections:

- Generating document reports
- Introducing report templates
- Generating HTML content
- Using charts

- Creating custom SQL queries
- The copy-and-paste approach

As usual, we will first look at what you will need to complete this chapter.

Technical requirements

Of course, you will need **Sparx Systems' Enterprise Architect** installed on your computer. We have included the report templates used in this chapter in the repository for this chapter. You can access the repository at the following GitHub address: `https://github.com/PacktPublishing/Practical-Model-Driven-Enterprise-Architecture/blob/main/Chapter11/EA%20Repository.eapx`.

You will need Microsoft Word, or a similar word processor, installed on your system to view and edit the documents generated by Sparx. Familiarity with Word is helpful but not altogether necessary. Sparx can generate documents in Microsoft Word format or **Rich Text Format** (**RTF**) for editing in other tools.

You will also need a web browser installed on your system to view and navigate the HTML pages generated by Sparx. Familiarity with HTML is helpful but not absolutely necessary.

Familiarity with **Structured Query Language** (**SQL**) is a plus for one section of this chapter. If you're not familiar with SQL, simply skip the section titled *Creating custom SQL queries*.

If you've never worked with report generators in the past, the concept of building templates may be new to you. We will cover templates in detail in this chapter, as it is key to understanding the publishing function. As with other report generators, understanding the underlying data structure is beneficial. Detailed coverage of the Sparx repository data structure is beyond the scope of this book; however, we will point you to resources that can be of great help in this regard.

Let's get started.

Generating document reports

The process for generating reports in Sparx is fairly straightforward:

1. You can simply identify the scope of the model elements to be included by selecting a package from the project browser, and then navigate to **Publish** > **Model Reports** > **Report Builder** and select **Generate Documentation**.

The following screenshot illustrates this navigation:

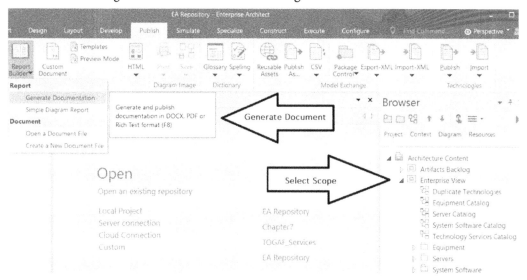

Figure 11.1 – Generating a report

When this action is taken, the following dialog appears:

Figure 11.2 – Generating the documentation dialog

As you can see, there are a lot of options in this dialog. For now, we will focus on the two that are required, the name and location of the output file and the report template to use:

- **Output to File**: This field lets you select a suitable location for the output file. The output file type indicates the type of file to be generated. In our example, the filename takes on the folder name from the project browser, and the `.docx` file type indicates that a Microsoft Word document will be created.

- **Template**: This field lets you select a report template. For our purposes, select the system-supplied template called **Model Report**.

RTF Files

Sparx supports the generation of `.rtf` files. **RTF** stands for **Rich Text Format** and is an early standard for sharing files across platforms. While Microsoft Word can edit both `.rtf` and `.docx` files, `.rtf` files tend to be much larger than their native Microsoft Word counterparts. You may also run into formatting options that Microsoft Word does not directly support. Still, `.rtf` files may be a good option when using tools other than Microsoft Word.

---

2.   Next, check the box for **View Document on Completion**.

3.   Now, press the **Generate** button.

As the report is created, the progress bar is updated. When it's complete, the document will be opened in Microsoft Word.

The following figure is an excerpt of the generated report:

Model Report   21 April, 2022

Enterprise View

Package in package 'Architecture Content'

Enterprise View
Version 1.0 Phase 1.0 Proposed
Joe & Robin Williams created on 9/20/2021. Last modified 9/20/2021

Duplicate Technologies diagram

Technology Layer diagram in package 'Enterprise View'

Duplicate Technologies
Version 1.0
Joe & Robin Williams created on 9/28/2021. Last modified 11/15/2021

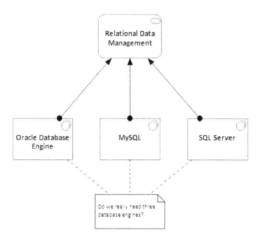

Figure 1: Duplicate Technologies

Figure 11.3 – A partial report output

If you were following along using the provided repository for this chapter, you may have
noticed that it took a long time to generate this report. You also may have noticed that
the report is almost 200 pages long. That's because we selected a high-level package with
a substantial amount of content. This is an important observation for organizing and
maintaining your repository going forward. You should always consider how you will use
your information.

We will review these and other approaches for organizing your repository in the next chapter. Another reason this report is so large is because of the large amount of information to be included for each model element in the report, as specified by the template.

Now that we have an idea of how to generate a report, let's turn our attention to report templates.

Introducing report templates

To better understand report templates, it's best to have a conceptual understanding of the Sparx repository structure. A Sparx repository is a collection of model elements. We place these elements within packages and represent them in one or more diagrams.

Elements, packages, and diagrams have various attributes. Attributes include things such as requirements, files, issues, constraints, or rules. Elements often have relationships or connectors to other elements. There are relationships that you create in a diagram by dragging a connector from one source element to another target element.

There are also parent-child relationships. For example, a package may contain or be a parent to many elements, diagrams, and other packages. An element may be nested within another parent element. Elements, attributes, connectors, diagrams, and packages all contain fields such as names, notes, and dates.

Report templates provide the specification of precisely what type of information to place on the report, where to place it, and how to format it. The information in a template includes elements, attributes, connectors, and diagrams, along with fixed text and formatting information.

We include this information in the template by selecting specific report sections and then specific fields for each section. Report sections are sets of opening and closing tags, much like what you see in an HTML document. These report sections have a hierarchy based on their relationships with each other. This all should become clearer as we view the template builder within Sparx. Let's go through the process of creating a simple report template from scratch:

1. Open the model repository for this chapter and navigate to **Publish** > **Model** > **Reports** > **Templates** options.

   A new blank template opens, as shown in the following figure:

Figure 11.4 – A new report template

   The panel on the left is where we specify the report sections to include in the template.

2. Expand the tree section labeled **Package** and check the box next to the **Element** section.

When you do this, Sparx places the appropriate opening and closing tags on the template panel to the right, as illustrated in the following figure:

Figure 11.5 – The section tags on the template

As you can see, the highest-level tag is the `package>` tag. Nested within the `package` tag is the `element>` tag. Nested within the `element` tag is the `child elements>` tag. At the bottom of the template are the complementing closing tags for each of the opening tags in reverse order, `<child elements`, `<element`, and `<package`.

Note that the **Package** section was checked and its tags included in the template for us when we checked the **Element** section. Elements must be nested within a package. The **child elements** section is also checked by default. If you expand the **Element** section in the **Sections** panel, you will see a large list of report sections that represent the attributes of an element.

Once we specify the report sections, we need to identify the fields to include for each section. Sparx has provided a hint on how to do this by including fixed text within the **Package** and **Element** sections.

3. Where it says [**right-click-to-insert-Package-field(s)**], remove this text and replace it with `This package is named ''`. Note the single quotation marks at the end of the previous sentence.

4. Place your cursor in between the single quotation marks and right-click. Sparx will present you with a context menu to select the fields available in the package report section.

5. From the context menu, select the **Name** field.

6. Select the entire line and set the format to **Arial**, **18**, and **underlined**.

7. Repeat these steps for the **Element** section. Indent and set it to **Arial** and **14**. Your template should look something like the following:

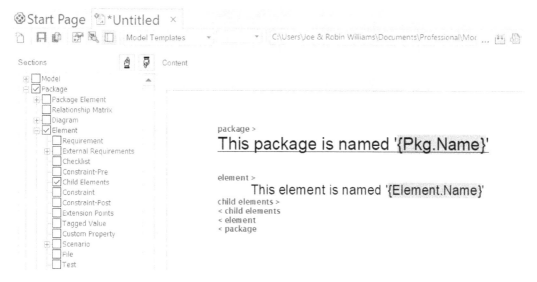

Figure 11.6 – My First Template

Save the new template and name it My First Template.

8. In the project browser, navigate to **Enterprise View** and select **Equipment**.

9. Press the *F8* key to load the **Generate Documentation** dialog.

10. Select **My First Template** as the template name. It should be located under the **Custom Templates** section.

11. Press the **Generate** key. The report is created in the specified output area.

If you had the **View Documentation on Completion** checkbox checked, Microsoft Word should open the output report, which should look something like the following:

This package is named 'Equipment'

> This element is named 'Cisco Router 3850'
> This element is named 'Cisco Router 3850'
> This element is named 'Cisco Router 3850'
> This element is named 'Cisco Router 3850'
> This element is named 'Cisco Switch 2960'
> This element is named 'Cisco Switch 2960'
> This element is named 'Cisco Switch 2960'
> This element is named 'Cisco Switch 3850'
> This element is named 'Cisco Switch 3850'
> This element is named 'Cisco Switch 3850'
> This element is named 'Dell Firewall SonicWall Nsa 2650'
> This element is named 'HP Rack HPE G2 Enterprise Series'
> This element is named 'HP Rack HPE G2 Enterprise Series'
> This element is named 'HP Rack HPE G2 Enterprise Series'
> This element is named 'IBM Power Controller 900942a'
> This element is named 'IBM Rack 32U'
> This element is named 'IBM Rack 32U'
> This element is named 'IBM Rack 32U'
> This element is named 'IBM Router 2210 nways'

Figure 11.7 – The My First Template results

In another example, we will show you how to copy an existing template. For this example, we will use the **DuplicateTechnologies** template created in *Chapter 7*, *Enterprise-Level Technology Architecture Models*. We will save the template to a new template group for easier reference. The steps are as follows:

1. From the **Template Editor** tab, press the **New template** button ( ), and the following dialog opens:

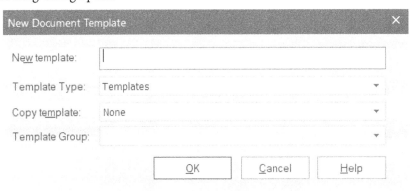

Figure 11.8 – The New Document Template dialog

2. Name the new template My Second Template.

3. In the **Copy template** field, select **DuplicateTechnologies** from the drop-down list.

4. In the **Template Group** field, enter Practical Model Driven Enterprise Architecture.

5. Press **OK**, and the new template opens in the **Template Editor** tab.

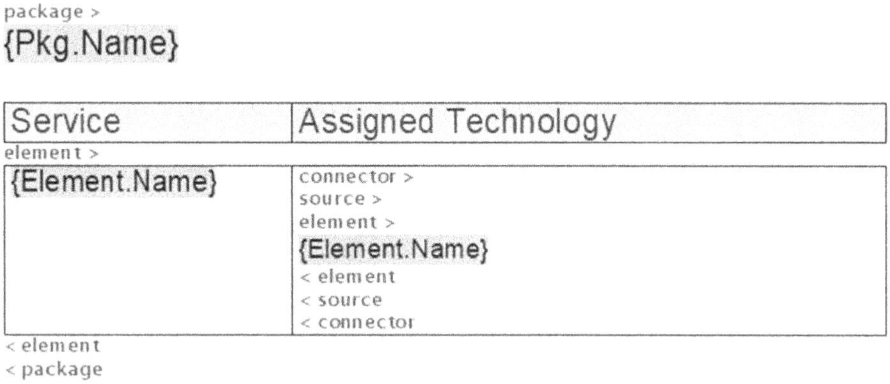

package >
{Pkg.Name}

Service	Assigned Technology
element >	
{Element.Name}	connector > source > element > **{Element.Name}** < element < source < connector

< element
< package

Figure 11.9 – My Second Template

Logically, this template behaves as depicted in the following pseudo-code:

```
For each package> selected
    print the package name {Pkg.Name}.
    Print a fixed text table header
    For each element> in package>
        print the {Element.Name} in table column 1
        For each connector> attached to element>
            look up the source> element>
            Print the source> element> {Element.Name} in
table column 2
```

Hopefully, this gives you a basic idea of how report templates work. Don't be misled by these simplistic examples, however. We have seen absolutely beautiful reports generated by Sparx. Unfortunately, a more detailed review of complex reporting templates is beyond the scope of this book. To refine your knowledge of report publishing, we suggest the following:

- Play with this new template by adding your own report sections and fields. Explore the available fields within each report section. This is also an excellent way to discover Sparx features that you may not have been aware of. Specifically, try including the report sections for child elements and child packages. Regenerate the report to see how the report content changes.

- There may be cases where the elements you wish to report on are not located within a single package. They may be scattered across multiple packages in the repository. In such cases, you can allow a model diagram to control the elements included in a report by including the **Element** subsection, nested within the **Diagram** section of the template.

- Explore the system templates provided by Sparx. You can open them in **Template Editor**. You won't be able to modify them, but you can copy them into new templates and modify the copy.

- Explore the Sparx help system, specifically the section titled *Model Publishing*. It's an excellent resource: `https://sparxsystems.com/ enterprise_architect_user_guide/15.2/model_publishing/ documentingprojects.html`.

Future-proofing your templates

As your repository grows, so will your collection of custom report templates. It's not unusual for a collection of templates to become unwieldy over time, making it difficult to find the specific template you need. To combat this effect, we recommend the following:

- Pay attention to the template names. Use descriptive names that include how a given template differs from others.

- When creating a new template, assign it to a specific template group. This will allow you to organize your templates and help narrow your search when the time comes to use the template. You can browse template groups from the **Resources** browser:

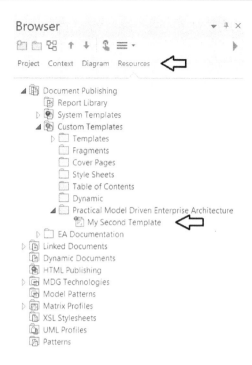

Figure 11.10 – Template groups in the Resources browser

- Look into the use of **template fragments** to format areas of reports that can be reused.

We will explore template fragments in the next section.

Template fragments

Commonly formatted sections of a template can be defined in a template fragment, which, in turn, may be included in multiple report templates. In *Chapter 7, Enterprise-Level Technology Architecture Models*, we used the `DuplicateTechnologies` template to identify potential duplicate technologies by listing technology services and all of the technologies assigned to each service.

However, as we can see from *Figure 11.9*, the only aspect of this template that is specific to technology services or technologies is the fixed text table header. Indeed, we can use this template to report on elements in any package and the elements connected to those elements. The only thing we need to change is the table header row. Everything between the `element>` tags can go into a template fragment. The Sparx Enterprise Architect User Guide provides some guidance on the use of fragments: `https://sparxsystems.com/enterprise_architect_user_guide/15.2/model_publishing/rtf_template_fragments.html`.

In the repository for this chapter, we've included another template called **My Third Template**, which includes a template fragment called **My First Fragment**. It produces the same results as **My Second Template**.

Now, let's look at other features of the report template.

Including diagrams in your report

Including diagrams in your report is straightforward. You need to decide whether you want to include diagrams at the package level or diagrams nested within an element. The **Diagrams Demo** template in this chapter's repository shows how to include a diagram image at the package level. The diagram is centered on the page and includes a figure number and diagram name:

Figure 11.11 – The Diagram Demo template

Now, let's look at the various means of getting just the content you're looking for in a document.

Document options

There are two ways to specify options for a document – through the **Generate Documentation** dialog or the **Document Options** dialog. The main difference between these dialogs is that the options selected from the **Generate Documentation** dialog are transitory.

They are, in effect, only for the current report generation. The options specified in the **Document Options** dialog are saved with the template and are applied each time the template is invoked. You can access the **Generate Documentation** dialog anytime by pressing the *F8* key or by navigating to **Publish** > **Model Reports** > **Report Builder** > **Generate Documentation**.

You can access the **Document Options** dialog by pressing the document options button from the ribbon on the **Template Editor** tab, as indicated in the following screenshot:

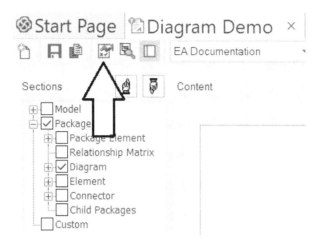

Figure 11.12 – The document options button

The effects of the specific options are self-explanatory, so we won't go over them all. Three types of filters are available:

- **Exclude Filters** allows us to exclude selected connector and diagram types, or exclude the details for the selected attributes in templates that would otherwise include those details.

- **Element Filters** allows us to specify the criteria for the elements we want to include in the document.

- **Other Filters** allows us to specify only the attributes to include in the document.

The element filters and other filters both include two checkboxes, one to add the filter to the document, and another called **Required** to indicate that the attribute in question must be present for the object to be included in the report.

There are a few menu options available in the **Generate Document** dialog that are not in the **Document Options** dialog:

- **Suppress Sections** allows us to select entire sections to be left out of the generated document.

- **Project Constants** allows us to define custom fields with fixed values to include in report templates. This can be used for anything from copyright notices to client names.

- **Word Substitutions** lets us define substitutes for certain words or phrases in a document. This may be useful for organizations that employ specific taxonomies in their model.

Excluding packages and diagrams from a document

You may find it necessary to exclude certain packages or diagrams from a report. This can be done as follows:

- Packages can be excluded by selecting a package and navigating to **Design** > **Model** > **Manage** > **Report Options**.

- Diagrams can also be excluded. From a diagram's **Properties** dialog, select **Diagram** and check the box labeled **Exclude Image from Documentation**.

Creating more complex documents

The Sparx document generation capabilities don't end with what we've covered so far, but comprehensive coverage of all of the capabilities and possibilities would require a book of its own. One capability that's worth mentioning, however, is what Sparx refers to as **virtual documents**.

Using this capability allows us to define complex documents containing sections from multiple areas of the repository independent of the repository structure and each using different templates. Standard fixed-text sections can be combined and mixed with variable sections. The virtual document can contain a cover page and a table of content, and use a standard style sheet. More information on virtual documents can be found in the Sparx user guide at the following address: `https://sparxsystems.com/enterprise_architect_user_guide/15.2/model_publishing/virtualdocuments.html`.

Hopefully, we've provided you with the fundamentals of document report generation. One effect that you may notice is that as your repository grows, so does the number of report templates and the size of their output. This is not a bad thing, as pulling information from your repository is its primary benefit. One disadvantage of document reports, however, is that as they are regenerated, they are replaced in their entirety, and they require a sequential access method. Now, we will turn our attention to a form of publishing that addresses this and other issues – web publishing.

Generating HTML content

The content of your repository can be made available for web access in one of three ways:

- **WebEA**
- **Prolaborate**
- Generating static web pages in HTML

WebEA is a feature of the **Sparx Systems Pro Cloud Server**. The Pro Cloud Server and Prolaborate are products provided by Sparx Systems that require separate licenses. While those products have great features, they are beyond the scope of this book. We will focus here on the static HTML pages that you can generate right from the product you have installed on your system.

By *static*, we mean that the information in HTML form is a copy of what is in your repository. When you change the repository, you must regenerate the HTML pages. Like the document report generator, you can select subsets of your repository to make them available via HTML. Unlike the document report generator, that subset will be fully navigable. The web pages generated contain a mixture of HTML, JavaScript, CSS, images, and similar constructs. All page references are relative, and all the necessary files are generated to allow you to simply drop the entire package structure into your web server's content area without modification. The only thing you may wish to change is the logo image that's displayed at the top of the page.

During the process of viewing and testing the generated web content, you will likely want to display the pages from your local system. Web browsers often don't like to access local files, so you may have to do some configuration in your web browser. We like using **Chrome** to test our page, but to do that, we must start Chrome from the command line with the following command-line option:

```
chrome.exe --allow-file-access-from-files
```

To configure your browser to view local files, see the section named *Browser Behavior* on the following Sparx user guide web page: https://sparxsystems.com/ enterprise_architect_user_guide/15.2/model_publishing/ htmlreport.html.

Once again, we will use this chapter's repository to demonstrate the generation of HTML:

1. Open the repository for this chapter.

2. In the project browser window, select and expand the view called **Enterprise View**.

3. Open the **Publish as HTML** dialog by navigating to **Publish** > **Model Reports** > **HTML** > **Standard HTML Report**. The following screenshot shows the dialog:

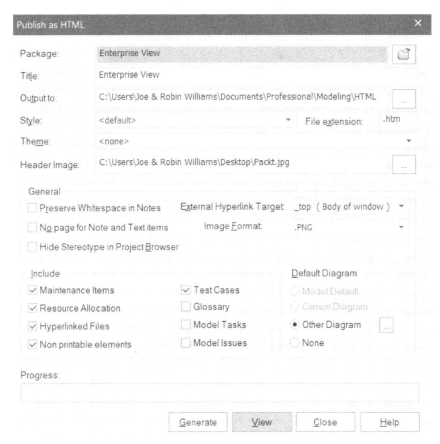

Figure 11.13 – The Publish as HTML dialog

4. Select an appropriate output path.

5. Optionally, select an image to be displayed on the header. We've selected the Packt Publishing logo.

6. In the **Default Diagram** section, select **Other Diagram** and navigate to a diagram to be displayed on entry to the website.

7. Press the **Generate** button. Sparx updates the progress bar and notifies us when complete.

8. Press the **View** button to view the output in your default browser.

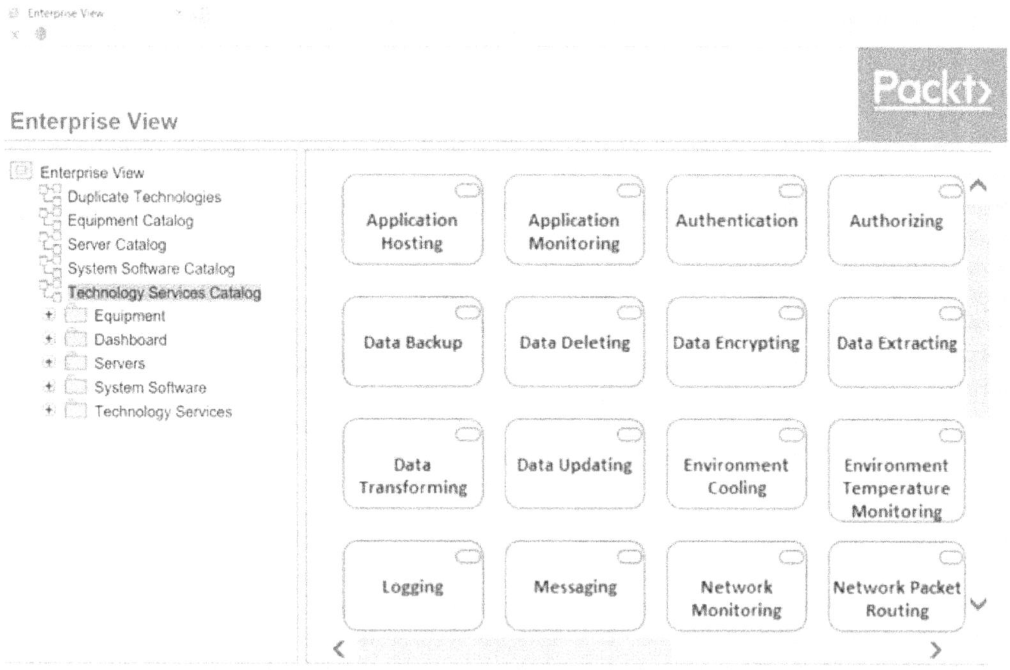

Figure 11.14 – The HTML report view

You can navigate through the web report by selecting a diagram or element from the navigation panel on the left, or by clicking on an element in the displayed diagram.

The contents of the output folder should look like the following:

css   EARoot   files   images   js   blank.htm   index.htm   toc.htm

Figure 11.15 – The content of the HTML report output folder

Point your favorite web server to this folder or place the folder in the server's context area, and you've got yourself a published model. That's all there is to it. Of course, you should probably notify your stakeholders when this occurs so that they know to access it. It's not uncommon to publish such content at regular intervals so that stakeholders can see the progress you've made in the model.

Another way to show progress is by displaying status, trends, and statistics through the use of graphs and charts. That is the subject of our next section.

Using charts

Charts and graphs are an excellent way of presenting another view of your enterprise. They can be used to report on the progress of a particular project or to illustrate the state of a particular aspect of the enterprise. Whatever your need, chances are that Sparx has the charting ability to satisfy it. Reviewing all of the charting capabilities of Sparx is beyond the scope of this book; however, we would like to show you some basics and how to get started with charting. We will do this by creating a simple dashboard in the Sparx repository. Let's get started:

1. If it's not already open, open the repository for this chapter. Select **Enterprise View** from the project browser and create a new folder called `Dashboard`.

2. With the new **Dashboard** folder selected, press the **New Diagram** button from the browser panel's ribbon.

3. From the **New Diagram** dialog, select the **Extended** perspective and select the **Dashboard** diagram type, as shown here:

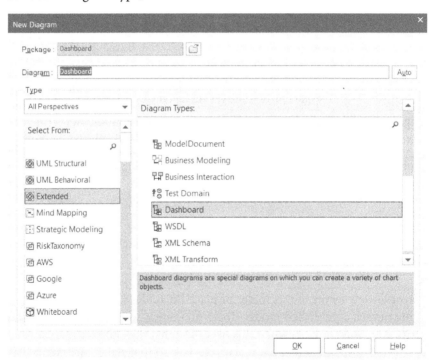

Figure 11.16 – The New Diagram dialog

4. A new, empty diagram is opened, and the diagram toolbox should look like the following:

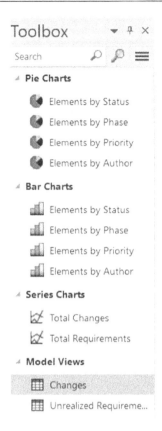

Figure 11.17 – The Dashboard toolbox

5. Under **Pie Charts**, drag the **Elements by Status** object onto the diagram. Sparx opens the **Find Package** dialog, allowing us to select the package on which to report:

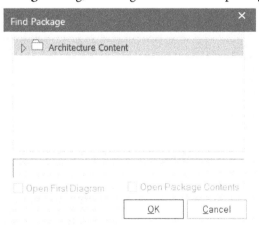

Figure 11.18 – The Find Package dialog

6. Expand the package to reveal the sub-packages and select the **ABC Trading Service** package nested within the **Enterprise View > Technology Services** package hierarchy.

7. Press **OK**.

Sparx places a pie chart on the diagram and immediately opens the following diagram:

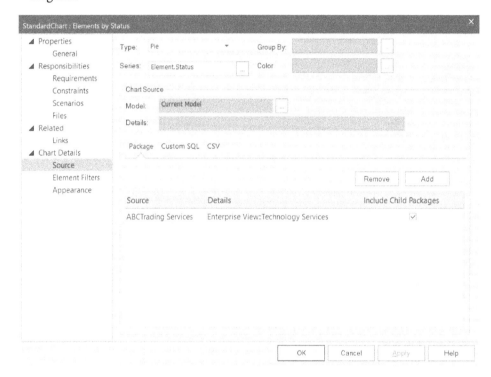

Figure 11.19 – The StandardChart : Elements by Status dialog

8. Take some time to explore this dialog:

    ▪ Note that additional sources can be added to the chart.

    ▪ Sources can be from package selections, custom SQL queries, or .csv files. This means that you can pull in information from external sources.

    ▪ Click the button next to the **Model** field to see the other types of sources that this chart might accept.

    ▪ Data for this chart can come from sources external to Sparx.

    ▪ In the dialog panel on the left, under **Chart Details**, select **Appearance**. Note the various options for displaying labels in this chart.

9. Select the **General** properties and change the name of the chart to `Technology Services by Status`.

10. Press **OK**.

The diagram should look like the following figure:

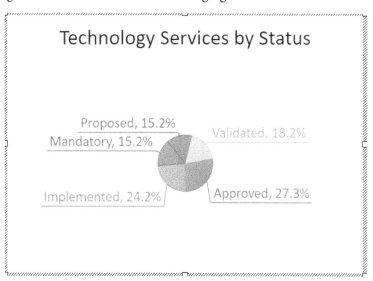

Figure 11.20 – Technology Services by Status

11. Move this chart to one side of the diagram. Using the same steps as previously described, drag the **Elements by Phase** bar chart from the diagram toolbox onto the diagram. Use the same package as the source. Name the chart `Technology Services by Phase`. The diagram should now look something like the following figure:

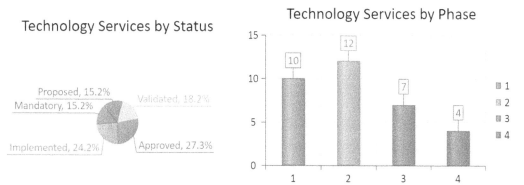

Figure 11.21 – The combined charts

12. Select one of the charts on the diagram and right-click. A context menu is displayed. One of the options on this menu is **Refresh Chart**. This must be done anytime the underlying data changes.

Where Did These Values Come From?

You may be wondering where the status and phase values came from. These are standard fields available on all elements in Sparx. We simply changed the default values with arbitrary values to better demonstrate the capabilities of these charts.

This diagram can be included in a document report or HTML page. It can be treated just like any other diagram. We can even make this the default diagram for the HTML report.

We've barely touched on the capabilities of the charting features. There are many more chart types and options available, but hopefully, this will get you started. You can learn more about them in the Sparx user guide at the following location: `https://sparxsystems.com/enterprise_architect_user_guide/15.2/model_publishing/charts.html`.

We saw that custom SQL queries were one of the options for defining the source of data for a chart. Let's take a little closer look at it next.

Creating custom SQL queries

If you're a keen observer, you'll have noticed that the subject of custom SQL queries has popped up in a couple of areas. SQL is an option for specifying the source used in document template fragments, charts, and custom queries. While the subject of SQL is well outside the scope of this book, for those advanced users who are familiar with SQL, we wanted to make you aware of the capability and provide you with some of the resources you'll need to work with it in Sparx.

In a nutshell, a custom SQL query allows us to provide the data expected for charts and template fragments by executing a SQL `SELECT` statement against the internal tables of the Sparx repository. The SQL statement must return each of the fields expected by the template fragment or chart. This is ideal for situations where the data we're looking for is not neatly contained within a single package or element type. You can learn more about using custom SQL queries in a chart from the following link: `https://sparxsystems.com/enterprise_architect_user_guide/15.2/model_publishing/chart_query.html`.

The problem with the custom SQL query approach is that the Sparx internal tables are not easily known to us. Sparx doesn't advertise its internal table structure. You won't find much coverage of this subject in the Sparx user guide. This is somewhat understandable, as they need the flexibility to change that structure at will to provide more and better functionality in future releases.

If we were ambitious enough, I'm sure we could do some poking around and discover a few interesting tidbits about the internal table structure. But we're not! After all, why should we go through all of that trouble when it's already been done for us? Thomas Kilian has produced the publication, *Inside Enterprise Architect – Querying EA's Database*, available at `http://leanpub.com/InsideEA`, in which he lists all of the Sparx tables that he's discovered and identified their content and use.

If you're interested in playing around with this feature, the best way to test SQL queries is by using the SQL Scratch Pad in the search facility in Sparx. Here's a sample navigation recipe:

1. Open the search facility – **Explore** > **Search** > **Model**.
2. Press the new search button on the **Search** tab's ribbon.

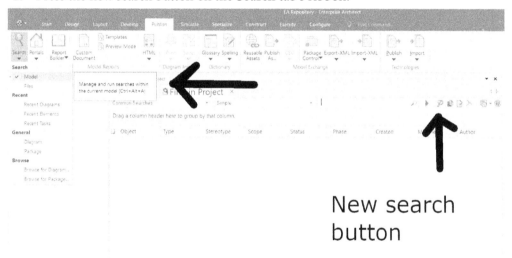

Figure 11.22 – Opening the search facility

3. In the resulting **New Search** dialog, provide a name such as `Test` and select the **SQL Edit** radio button:

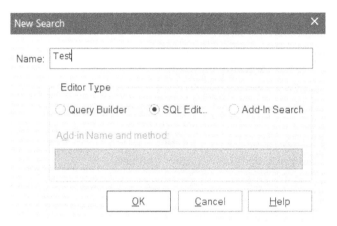

Figure 11.23 – The New Search dialog

4. Press the **OK** button. In the upper panel of the resulting search tab, select **SQL Scratchpad**.

5. In the **SQL Scratch Pad** editor, enter a SQL query such as the following:

```
select status, name from t_Object where stereotype =
"ArchiMate_TechnologyService"
```

6. On the **Search** tab's ribbon, press the run button (the green arrow) next to the new search button. The search results panel should be populated with multiple rows of status and name fields.

7. On the search results panel, drag the status column header onto the grouping panel just above the search results panel. Your results should look something like the following screenshot:

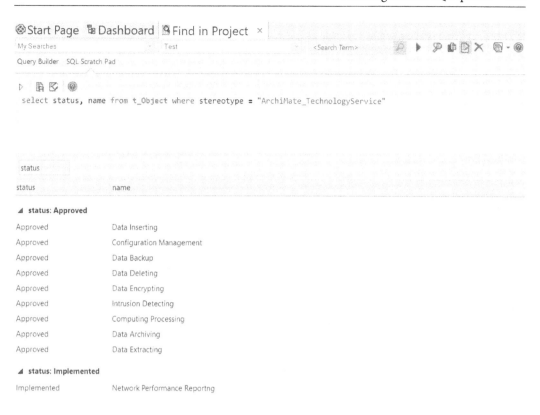

Figure 11.24 – The SQL query results grouped by status

The **SQL Scratch Pad** editor uses what Sparx refers to as the **Common Code Editor**. It has IntelliSense features such as code completion. For example, when entering the table name followed by a period, a drop-down list of the available fields from that table will appear. Of course, another way to discover the available fields is to enter the following statement:

```
select * from t_Object
```

Once you have tested your query and are satisfied that it returns the appropriate results, you can save the query, or copy the SQL statement and paste it into the appropriate area of the document fragment or chart builder dialogs.

SQL queries are not for the faint of heart. This is definitely a subject for advanced users. Now, let's move on to a simpler subject that everyone can take advantage of and that is used most often – the simple copy-and-paste method of publishing.

The copy-and-paste approach

We would be remiss if we didn't mention the simplest publishing mechanism of all, copy-and-paste. For simple, one-off communications and documents, this is indeed the simplest means of conveying model diagrams. Because most are already familiar with this approach, we won't spend much time on this subject. There are just a couple of things to go over for new Sparx users.

There are two ways to select and copy diagrams. You can use the menus by navigating to the following:

- **Layout** > **Diagram** > **Select** > **Select All Elements**
- **Layout** > **Diagram** > **Select** > **Copy Selected Element(s)**

You can also simply drag your mouse cursor over the elements you wish to copy and press *Ctrl + C.*

When it comes to pasting the clipboard contents, you may get a bit of a surprise. Sparx (**Enterprise Architect version 15.2** and earlier) places a border around all diagrams, copied using the aforementioned means. Here's an example of the default border:

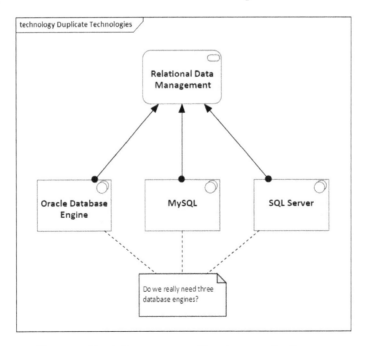

Figure 11.25 – A Sparx-generated border around a diagram

If you don't mind this border, then you're in business. Otherwise, you'll need to paste the image into a tool such as **Microsoft Paint** and crop out the portions of the diagram you need before pasting it into a document. This is how all of the diagram images in this book were produced.

Summary

Next to gathering information, publishing is the most important capability of your enterprise repository. There wouldn't be much use for a repository if it didn't have such capabilities. Fortunately for us, Sparx has a rich set of publishing features.

We've barely brushed the surface of what this tool can do. However, we're afraid that delving any deeper into the various features and uses of this tool at this time would lose all but the most adamant and determined reader. There's a better way to learn, which is to experience it for yourself. Go forth and model!

Within the chapters of this book, we have endeavored to teach you one of the only modeling languages specifically intended for the enterprise architect, ArchiMate®, using one of the most popular, feature-rich, and affordable tools of the enterprise architect, Sparx Systems' Enterprise Architect. We hope we have succeeded.

The best architectural decisions are based on the best information. We believe that it is the primary job of the enterprise architect to make such information available. Whether your organization makes architectural decisions by consensus, committee, decree, or yourself, you now have the tools necessary to make those decisions correctly. You have the tools necessary to design, generate, structure, organize, and clearly communicate information about your enterprise. You have the tools to accurately identify and communicate the current state of your enterprise, the options available for moving forward, the reasons for selecting one path over another, and an understanding of the consequences of selecting the wrong path. With such information and knowledge, how could you not succeed?

Index

Packt.com

Subscribe to our online digital library for full access to over 7,000 books and videos, as well as industry leading tools to help you plan your personal development and advance your career. For more information, please visit our website.

Why subscribe?

- Spend less time learning and more time coding with practical eBooks and Videos from over 4,000 industry professionals

- Improve your learning with Skill Plans built especially for you

- Get a free eBook or video every month

- Fully searchable for easy access to vital information

- Copy and paste, print, and bookmark content

Did you know that Packt offers eBook versions of every book published, with PDF and ePub files available? You can upgrade to the eBook version at packt.com and as a print book customer, you are entitled to a discount on the eBook copy. Get in touch with us at customercare@packtpub.com for more details.

At www.packt.com, you can also read a collection of free technical articles, sign up for a range of free newsletters, and receive exclusive discounts and offers on Packt books and eBooks.

Other Books You May Enjoy

If you enjoyed this book, you may be interested in these other books by Packt:

Solutions Architect's Handbook - Second Edition

Saurabh Shrivastava, Neelanjali Srivastav

ISBN: 978-1-80181-661-8

- Explore the various roles of a solutions architect in the enterprise landscape
- Implement key design principles and patterns to build high-performance, cost-effective solutions
- Choose the best strategies to secure your architectures and increase their availability
- Modernize legacy applications with the help of cloud integration
- Understand how big data processing, machine learning, and IoT fit into modern architecture
- Integrate a DevOps mindset to promote collaboration, increase operational efficiency, and streamline production

Oracle Autonomous Database in Enterprise Architecture

Bal Mukund Sharma, Krishnakumar Mohanram, Rashmi Panda

ISBN: 978-1-80107-224-3

- Explore migration methods available for Autonomous databases, using both online and offline methods

- Discover how to create standby databases, RTO and RPO objectives, and Autonomous Data Guard operations

- Become well-versed with automatic and manual backups available in ADB

- Implement best practices relating to network, security, and IAM policies

- Manage database performance and log management in ADB

- Understand how to perform data masking and manage encryption keys in OCI's Autonomous databases

Packt is searching for authors like you

If you're interested in becoming an author for Packt, please visit `authors.packtpub.com` and apply today. We have worked with thousands of developers and tech professionals, just like you, to help them share their insight with the global tech community. You can make a general application, apply for a specific hot topic that we are recruiting an author for, or submit your own idea.

Share Your Thoughts

Now you've finished *Practical Model-Driven Enterprise Architecture*, we'd love to hear your thoughts! Scan the QR code below to go straight to the Amazon review page for this book and share your feedback or leave a review on the site that you purchased it from.

`https://packt.link/r/1-801-07616-2`

Your review is important to us and the tech community and will help us make sure we're delivering excellent quality content.

www.ingramcontent.com/pod-product-compliance
Lightning Source LLC
Chambersburg PA
CBHW081504050326
40690CB00015B/2923